OSI CONFORMANCE TESTING METHODOLOGY AND TTCN

OSI CONFORMANCE TESTING METHODOLOGY AND TTCN

Bernd Baumgarten
Alfred Giessler

Gesellschaft für Mathematik und
Datenverarbeitung mbH (GMD)
Darmstadt, Germany

1994
ELSEVIER
Amsterdam – Lausanne – New York – Oxford – Shannon – Tokyo

ELSEVIER SCIENCE B.V.
Sara Burgerhartstraat 25
P.O. Box 211, 1000 AE Amsterdam, The Netherlands

ISBN: 0 444 89712 7

This book is printed on acid-free paper.

Printed in The Netherlands.

FOREWORD

The seven part International Standard, ISO/IEC 9646, the OSI Conformance Testing Methodology and Framework, has been 10 years in the making. It is undoubtedly the most comprehensive conformance testing methodology standard in any branch of Information Technology. Now there is an equally comprehensive book on the subject, making the ideas accessible to a much wider audience.

As the ISO/IEC rapporteur for the subject from the beginning in 1983, I am well aware of the contributions that different people have made to the work. The authors of this book have both been significant contributors and hence can write with real authority on the subject. Alfred Giessler has been involved from the beginning, hosting the very first rapporteur group meeting in February 1984. Bernd Baumgarten, on the other hand, is one of the select few who had a major influence on the design of the test specification language, TTCN. He contributed not only to the first edition of the TTCN standard, but also made significant contributions to the first amendment which defines the concurrency extensions.

ISO/IEC 9646 is now a very widely used standard. It is used by protocol and profile specifiers to help them specify their conformance requirements unambiguously and to provide proformas for suppliers to fill in to clarify which capabilities their products support. It is used by test specifiers to define how to specify OSI conformance tests. It is used to give guidance to test system developers. It is also used by test laboratories and the relevant accreditation bodies to define how an OSI test service should be operated. Finally, it is used by other testing methodology specifiers to serve as a guide on how to write testing methodology standards.

Thus, even though OSI has lost much of its earlier popularity, the conformance testing methodology continues to be of relevance to a wide variety of organisations.

This book is likely to be regarded as the definitive text book on the subject for some years to come.

Dr. Dave Rayner

ISO/IEC rapporteur for OSI conformance testing,
Chairman of the EWOS Expert Group on Conformance Testing,
Head of the Open Systems Group at the UK's National Physical Laboratory.

We dedicate this book to the memory of our friend Jean-Daniel Colas.

PREFACE

The goal of this book is to give an overview, a textbook, and a critical review of the Conformance Testing Methodology and Framework. This standard, ISO/IEC 9646, describes a methodology for testing implementations of Open System Interconnection (OSI) specifications. Our aim is to enable the readers of this book to understand exisiting test documents, especially test suites, and to write such documents on their own.

As to the book's role as an overview: most sections provide an outline of the main ideas in the first subsections. TTCN dynamics is summarized in the first Section of Chapter 4.

In the technical chapters, we present an in-depth treatment of the basic concepts, i.e. of test configurations, the syntax and operational semantics of the test notation TTCN, as well as the practical procedures in testing. In Chapter 5, we use a large coherent example, based on a simple protocol introduced particularly for the purpose, to demonstrate the various test related documents called for in the standard. A list of abbreviations and an extensive index support the lookup of the many interrelated notions. A tentative list of test related tools may be of help for those who want to work actively in the area.

Even though we certainly appreciate the achievements of the standard, our presentation is not uncritical, as we also identify problems and suggest preferable interpretations as well as improvements.

This book is not intended to replace the standard in any way. Neither were the authors able to condense, without loss of information, its seven parts to the 230 pages available for the technical Chapters 2 through 4, nor can the authors guarantee the total agreement of their text with the letter and intent of the standard. If you need an introduction to ISO/IEC 9646 or a critical review of it, then you may profit from reading this book. If you need the standard, then please use the standard.

The versions of the parts of the standard described in this paper, especially the second version of TTCN, were not yet published at the time of writing. Therefore, we based our treatise on preliminary versions. Since we could not be sure of the exact wording of the future publications, this entails certain inaccuracies of our presentation. We preferred this risk to the disadvantage of discussing versions that would soon be outdated. It is possible that some of the TTCN problems that this text identifies in the standard will already be solved in the forthcoming TTCN version. This text is a translated, corrected, updated, and extended version of [BAGI 92].

We are strongly indebted to Ron Winston, who suggested innumerous editorial improvements and contributed scores of technical comments and suggestions.

We profited very much from the clarifying and constructive discussions with the rapporteur, Dave Rayner, and the participants of the ISO/IEC 9646 editing meetings, the late Jean-Daniel Colas, Anthony Wiles, Ron Winston, Jan van Kroon, and Os Monkewich, to name but a few.

DIN meetings on conformance testing proved to be another source of ideas and interpretations; here, we had the pleasure to work, for example, with Otto Lang, Doris Berr, Jörg Burmeister, Michael Hillier, Udo Bär, Hans Zischler, Hartmut König, Jens Linke, Andreas Ulrich, Hans-Michel Staab, and Arno Reinshagen.

We are grateful for discussions with, and improvements suggested by, Eckart Raubold, Heinz-Jürgen Burkhardt, Christa Paule, and Helmut Wiland. We appreciate the stimulating discussions on conformance testing architecture that we had with Gregor von Bochmann, Dieter Hogrefe, and Reinhardt Gotzhein. Chu-Hwan Yim kindly provided us with information on ITU-T.

We thank ISO for the permission to reproduce the TTCN BNF rules. We present in Annex A a re-edited form of the preliminary version of October 1993, on which the BNF in the second version of TTCN will be based. We are grateful to the companies, university departments, and institutes who provided us with information on their test tools. We present these contributions, in condensed form, in Annex B.

The editors deserve our gratitude for their patience.

We will be thankful for any criticism, comment, or suggestion regarding the book.

July, 1994

Bernd Baumgarten
Alfred Giessler

CONTENTS

In this chapter, we will address some factors that influence OSI conformance testing.

The majority of OSI implementations are produced in the form of software that can be executed on various hardware platforms. In these cases, conformance testing is a particular form of software testing, though not very akin to the white box variant that is used in development testing.

OSI specifications is the generic term for OSI protocol, profile, information object, and transfer syntax standards. An *OSI implementation* is an implementation of an OSI specification. Protocols are the most important group of OSI specifications. Their issues are special cases of distributed systems issues. A *profile* is a dedicated version of a "stack of protocols," each of which is based upon the service provided by the one below it. A *transfer syntax* (also called a set of encoding rules) determines the bitstrings representing data objects in transmission. An *information object* is a "static" data object, as opposed to a "dynamic" protocol, which may include the operations that can be performed on it; examples are FTAM (file transfer, access, and management) document types and managed objects.

Particular problems have to be faced in distributed systems testing, and hence in protocol implementation testing. ISO and CCITT (now ITU-T) have invested considerable efforts in the standardization of an open systems conformance testing methodology, aimed at harmonizing tests of OSI implementations – a first step towards universally recognizable test results and certificates.

1.1 Software testing

ISO and CCITT/ITU-T have standardized a family of *Open Systems Interconnection* protocol standards and related recommendations. The role of these organizations is presented in the fourth section of this chapter. The standardized conformance testing methodology of OSI specifications is described in the international standard ISO/IEC JTC1: IS 9646 [ISO 9646]. As the testing of protocol implementations amounts to a special variant of software testing, the classical principles and techniques of software testing [BEIZ 90, MYER 91, PERR 83, SHOO 83] apply – with the appropriate modifications which we will discuss below. This holds true, in particular, with respect to the writing of test purpose catalogs.

We cannot really provide an overview of classical software testing in this chapter. We will, however, recall a few important aspects of this discipline, and discuss their relevance to OSI conformance testing methodology.

1.1.1 Black-box and white-box testing

Basically, the test philosophies of black-box testing and white-box testing are distinct, although a variety of intermediate and mixed forms ("grey-box") have been identified.

White-box testing means testing software for which the source code is laid open. In particular, the code's structure is known, both in the large (modules, procedures, data management) and in the small (decisions, jumps, loops etc.). Typical white-box test techniques are code inspection, walkthrough, design review, as well as methods relying on the analysis of data and control structures, flows and paths. Accessible program code may be manipulated for the purpose of testing, e.g. by temporarily inserting trace outputs, replacing modules, or even introducing intentional errors.

OSI conformance testing methodology was mainly introduced for the purpose of regulating and harmonizing *third party tests*, i.e. for situations where the source code is usually not available. (First party denotes the provider, second party the procurer of software, while third parties are any other involved parties with roles different from provider or procurer). Hence, testing in accordance with the OSI conformance testing methodology is generally restricted to *black-box testing*. OSI specifications are meant to be specifications of the class of correct external behaviors of the software to be tested, as realized by the *implementation under test* (*IUT*). Apart from descriptions of valid behavior, they neither contain nor directly imply any particular implementation details.

As the source code is left out of consideration, a completed test can only provide knowledge of *what* the IUT has done wrong, i.e. in which respect it behaved incorrectly. It cannot tell *where* in the source code the error occurred. Error localization beyond that which can be determined from the ordinary test campaign documentation, and the measures to be taken for error correction, are not subjects of [ISO 9646]. They remain the test client's concern.

Under these circumstances, white-box test techniques apply only to a very limited degree to the protocol testing scenario considered here. Yet, they are not totally irrelevant if we make the assumption that the program structure of the IUT's source code usually corresponds closely to the structure of the OSI specification. Further, for commercial reasons, one may ordinarily assume that no tricky features are built into the IUT which make it behave illegally just after the testing is over, such as selecting different choices depending on the date of the program execution.

The test cases to be undergone are well-determined and fixed in advance, but usually as a pre-determined function of the variant of the protocol declared to be implemented. The spectrum and the declaration of these variants, as well as the determination of the exact tests to be executed are explained in the second chapter. The set of test cases selected for the given protocol variant is called a selected *test suite*, and is expected to be passed completely by the IUT. Issues of test metric, like the probability of remaining errors (occurring per time unit or contained per lines of code) or the residual gain of testing (the probable rate of errors detected by further testing per time or monetary unit) are not treated in [ISO 9646]. OSI conformance tests are deterministic and not random.

The principle of pre-determined and publicly known test suites could theoretically invite the implementation of mere "test passers," which do not provide the full and correct set of protocol functions, but are just able to cross the hurdle of conformance testing. However, vendors would not be able to sell many copies of such a software, and, furthermore, standardized test suites are so voluminous and thorough that it could mean an extra effort to implement a mere test passer instead of a real protocol implementation.

1.1.2 Bottom-up and top-down testing

The notions of *bottom-up* and *top-down testing* refer to the stepwise decreasing or stepwise increasing granularity of the source code components being tested. As a part of white-box testing, strictly speaking, they do not apply to conformance testing.

In a figurative sense, however, applied to the hierarchical structures manifest in service and protocol standards, the same distinction can be made with respect to protocol conformance testing. In fact with the three categories proposed in [ISO 9646], basic interconnection tests, capability tests, and behavior tests (cf. Subsection 2.3.4), test suites follow in a certain sense the top-down orientation. This is also somewhat reflected in the hierarchical group structure provided for test cases within test suites. It should be noted, however, that single test cases remain explicitly independent units that can be executed in any order.

With respect to the hierarchy of OSI layers, implementations comprising a stack of protocols are usually tested layer by layer from bottom to top. Yet this layering is not directly related to the granularity of modules.

1.1.3 Error classes

In the strict sense, the differentiation of classes of errors, such as functional errors, structural errors (interface, operating system, software architecture, and resource problems), procedural errors (operational, arithmetical, initialization, logical, and control flow errors), data errors and syntax errors, refers to the source code. Analogously applied to the specification level, these types can be taken into consideration in the design of the test purpose catalogs. Data errors in particular play a significant role in protocol testing, usually inducing more test cases than the protocol logic.

1.1.4 Test phases

Test phases are often defined to correspond with phases of the software development process. We will not go into the details of any one of the multitude of software development phase, and software life cycle, models. Instead, let us merely mention module, integration, function, acceptance and installation testing, to name but a few typical test phases.

If we consider a protocol IUT as a module, OSI protocol tests are mainly concerned with module and functional testing. The IUT is usually not tested in the target environment where it is finally to be used; at least it is not integrated with the other software of the procurer. Instead, for the purpose of OSI testing it is considered as relatively isolated, directly interacting only with an underlying service provider and an operator or a tester component on top. Even if the IUT consists of a "stack" of protocol implementations, it is attempted to test them in relative isolation from one another. Acceptance testing is not a concern either, because principally the behavior of the IUT is prescribed by standards, and the final user's judgement is not considered. Implementation-specific IUT properties, such as the operator interface or the realization of the service primitives, are not investigated in OSI testing, because they are explicitly omitted in OSI specifications.

1.1.5 Testing techniques

Classical testing techniques are of varied relevance to OSI testing. Again, techniques which ordinarily demand the availability of source code can be applied only analogously with respect to structures of the IUT specification. An example of this is path testing, a class of testing techniques in which, roughly speaking, all program statements and all alternative choices are to be executed at least once. In the design of a test purpose catalog, this technique can be applied in a similar sense to protocol automata in the protocol specification. More specification-oriented methods, such as transaction flow testing and logic based testing, are directly applicable for the definition of test purposes and the design of the associated test cases.

From some perspectives, an IUT and its OSI specification can be viewed as two finite input-output-automata. Under certain very restrictive assumptions the existence of complete test suites can be shown. Passing a complete test suite amounts to a correctness proof of the IUT (under the given assumptions, which normally concern determinism and/or the number of states). The current, more realistic approach of using incomplete yet sufficiently thorough protocol tests is the basis for ongoing research on the definition and calculation of useful measures for the degree of coverage achieved by a test suite.

Robustness tests, i.e. tests of the capability of the IUT to handle invalid (e.g. incorrectly formatted) input are only considered in the limited cases explicitly mentioned in the protocol standard. These cases usually deal with reactions to invalid protocol data units or encodings, and never with incorrect service primitives. Any tests of (even very sensible) capabilities which are not explicitly stipulated in the protocol standards are considered as out of scope in [ISO 9646]. This includes robustness against any error not mentioned in the standard. The same exclusion holds for any kind of performance testing which would exceed the requirements specified in the standard.

1.1.6 Test realization

Test realization comprises aspects of the respective roles, the tools and the personnel involved in the process of testing, as well as the different activities and stages of planning, performance and documentation. In several sections of Chapter 2 we will describe how [ISO 9646] regulates the roles of test client and test laboratory as well as the planning and execution of tests themselves, including the selection of applicable test cases and the preparation of an executable test suite. Possible test configurations are listed, and minimal requirements of documentation are stated. It is the objective of national accreditation guidelines to ensure the qualification of testing personnel and the quality of test tools.

1.1.7 The relevance of a test

A test which has not been passed reveals one or more bugs in the investigated software. Though undesired from one point of view, such a test result has the definite advantage of yielding some concrete information about the IUT. Even though the rate of errors per line of code is usually higher in corrected code than in average original encoding, it is usually much lower than the rate in the lines which are identified as erroneous, and which are thereupon removed or replaced. Hence, failed tests sooner or later do lead to improved software.

On the other hand, a passed test gives the observer very little definite information. No error was discovered this time, but it is unknown how many remain undiscovered, or even whether a rerun of the same test would also be passed. As mentioned before, only under very particular circumstances, involving white box knowledge, can a complete test suite be determined, which, if passed, proves that the IUT is free of errors. The inherent complexity of communication protocols handicaps the search for complete test suites, even where they should exist. And the commonly found ambiguity, incompleteness and inconsistency of protocol specifications may render this search impossible. As a rule, the passing of a test is merely a statistical piece of evidence that the investigated software can now be trusted somewhat more than before the test. Usually, conformance testing and the ensuing corrections of the IUT will increase the probability that this IUT will be able to interwork in the expected way with other OSI implementations.

Due to the fundamental limits of testing, exacerbated by testing problems inherent in the distributed nature of the systems concerned, protocol testing requires

- a well-founded testing methodology, as well as
- test suites of high coverage, in particular with respect to "typical errors."

A set of test results will never be generally accepted unless the tests have been based on a generally recognized testing methodology and test suites. These are the subjects of past and ongoing standardization efforts. Accreditation requirements have already been mentioned in the discussion of test realization. The ultimate aim of these efforts is the granting of meaningful and widely recognized test certificates for protocol implementations.

1.2 Distributed systems

Distributed architectures arose in response to several needs:

- spatial distribution of system components dictated by the problem area, e.g. in the realm of communication systems;
- functional independence;
- higher processing speed, throughput, and reliability;
- sharing of common resources.

All these requirements suggest a system realization in the form of cooperating processes. These processes can run either physically separated on several processors or only logically separated on one processor. When they work with unrelated speeds and in relative independence from one another, they are called *concurrent*.

In *distributed systems* in the broader sense, processes run on separate processors, and mostly concurrently. Distributed systems in the narrow sense are only loosely coupled, i.e. by message exchange. Tightly coupled processes share some common memory.

1.2.1 Cooperation problems

Independently from the particular application case, concurrent processes cooperating for some common purpose are generally confronted with problems due to their distribution. Usually one attempts to solve these problems in advance, either by the provision of elementary coordination and communication services, on which the system is then based, or by the participating parties' general adherence to certain rules of behavior which guarantee a successful communication and cooperation. We will now take a look at some typical services or rule sets developed for this purpose.

The global consistence of set of the components' states is jeopardized if cooperative functions delegated to some local entity cannot be provided in the required degree. This danger is countered by methods of transaction control, e.g. by structuring and managing activities as possibly nested atomic actions, by repetition, reservation, or time stamp schemes, or by other measures to achieve fault tolerance.

In order to avoid unfairness or cyclic waiting situations (deadlocks) in the shared use of resources, well-known algorithms for shared resource administration are applied. Regardless of resource problems, the danger of deadlocks may be met by the use of deadlock avoidance or deadlock discovery schemes.

Methods for mutual exclusion prevent undesired interference between different processes in critical sections, such as write operations to common variables.

Another typical problem class for distributed systems is the achievement of global agreements, e.g. on common choices, on a consistent view of a body of data, or on the moment for a common transition to a new phase of the cooperation. These are frequently called consensus problems. Different additional constraints may be imposed, be it to achieve the agreement as efficiently as possible (fast or with few messages etc.) or in relative synchronism. In numerous publications either the correctness of some consensus algorithm or the non-existence of such an algorithm is proven, depending on the problem parameters. The latter comprise assumptions about the quality and quantity of the processes involved as well as the quality of the communication medium used.

These problem classes are not independent of one another. The solution of one problem can be a necessary or sufficient prerequisite for the solution of another problem. It can also be the source of new problems.

In the next subsection we take a look at testing problems in the context of distributed systems. Then we will concentrate on the topic most directly connected with that of OSI protocols, the communication problems of distributed systems and their mastering by means of communication protocols.

1.2.2 Testing distributed systems

Compared to the conditions in the local case discussed in the previous section, the testing of distributed systems raises some additional problems, in particular

- non-repeatability,
- state space explosion,
- non-existence of global views,
- non-observable parts, and the
- distribution of test systems.

In contrast with local deterministic software, systems involving concurrency are usually inherently non-deterministic, i.e. the results they produce are not fully repeatable, even with identical inputs. One reason for this is that both processing speeds in the local system components and transmission times of messages between these components may vary in different runs. Thus, time conditions may differ from one run to the next, and so will time-dependent reactions (such as timeout reactions and order-dependent message treatment) in system components.

Distributed systems have for combinatorial reasons usually a much larger number of global states and possible behaviors than purely local ones. This is called the "state space explosion" and complicates the analysis of the entire system and also the design of global behavior tests.

In contrast with the local case, there usually does not exist a global view of the complete system nor a tangible global state. Instead, all one can expect to observe is a mere collection of locally observable event sequences. Global system behavior may be hard to reconstruct from these local views, if not impossible.

Not all parts of the system, for example communication channels, are observable, much less controllable for the purpose of testing.

Test systems themselves have to be designed as distributed systems, in order to be able to handle the relevant set of observation points. Their components, as parts in the total test configuration, must each be realized correctly and must also cooperate properly. Thus, they are inevitably subject to the coordination problems discussed above.

1.3 Communication protocols

Communication problems in distributed systems can be grouped into two categories, depending on whether they are caused by

- the heterogeneity of the computers involved, along with the ensuing differences in communicative behavior, or,
- transmission errors, such as garbling, loss, reordering or duplication of messages, whether immediately recognizable or not.

The task of communication protocols is to provide for the efficient, free, and safe exchange of messages and signals. They achieve their purpose by realizing common strategies involving the request, composition, redundant encoding, sending, decoding, repetition, awaiting, reception, error recognition, storage, acknowledgement, etc. of data units.

In addition to these basic activities, performed mainly to overcome transmission errors, there are other technical and substantial aspects of protocols, such as:

- interface characteristics,
- synchronization,
- time limits,
- addressing,
- connection control,
- routing,
- composition and decomposition of data blocks (chaining, segmentation),
- composition and decomposition of data streams (multiplexing, demultiplexing),
- flow control,
- dialog modes (simplex, half-duplex, duplex) and sending rights,
- restoration of lost component or dialog states,
- data presentation,
- identification or anonymity,
- protection against all kinds of unauthorized access,
- network management.

All play an important role in the achievement of secure transmission and in the realiziation of higher communication services.

From this incomplete list of communication problems, activities, and characteristics one can easily infer that unstructured, monolithic protocols, which try to address all aspects of communication at the same time, can hardly be envisaged to succeed. For this reason, providers of communication systems, and even more so the national and international standardization committees concerned, soon conceived and/or adopted hierarchical communication architectures. In these, communication functions are grouped into modules, and these modules are ordered into a user-provider hierarchy. This view is nothing more than a transfer of the software-technological principle of hierarchically ordered virtual machines to the context of distributed systems.

1.3.1 The OSI Basic Reference Model

The Open Systems Interconnection (OSI) Basic Reference Model (BRM), described in 1984 in an ISO standard [ISO 7498], introduces a hypothetical layered structure for protocols, thereby providing a uniform frame of reference for the description of the external behavior of protocol instances. It abstracts from programming languages, operating systems, application and human user interfaces, and explicitly avoids placing any requirements on the actual internal structure of protocol implementations. The principle behind the hierarchical structure is illustrated in Figure 1.1.

A global service of layer N (the "(N)-service" for short) shall make its elements and service primitives (SP) available in different local systems at (N)-service access points (SAP). The standardized *service description* specifies the availability, use and effect of these service elements, but not their implementation or concrete formats of service "calls."

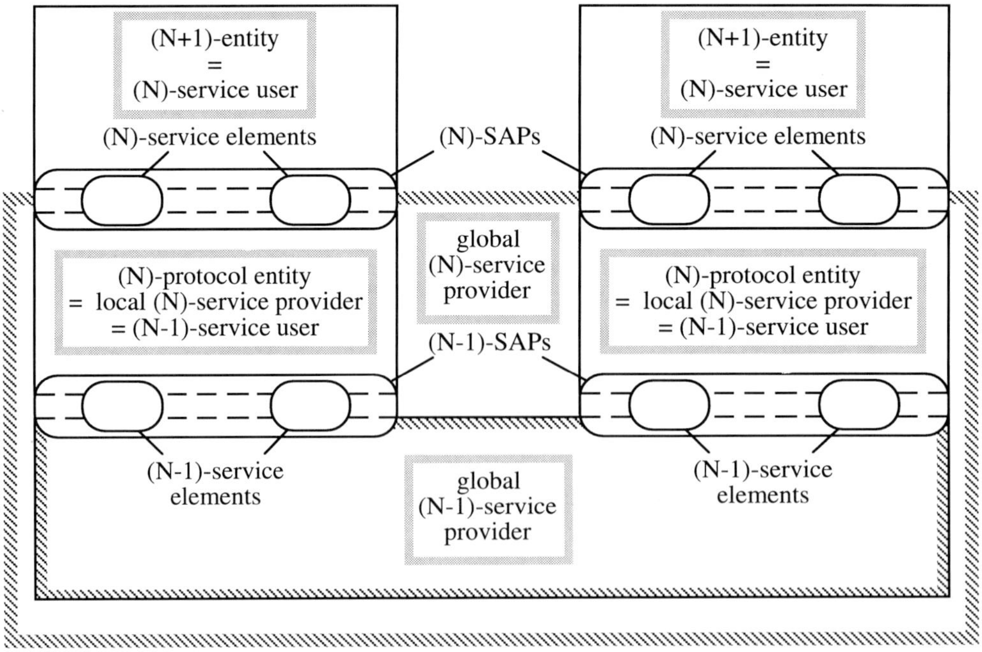

FIGURE 1.1. The principle of the service-protocol-hierarchy

Assume the (N)-service is not yet available, but an underlying (N-1)-service is. Then the (N)-service can be provided by two distributed (N)-protocol entities communicating and cooperating via (N-1)-service elements, in particular exchanging (N)-protocol data units (PDU). The description of the relevant protocol mechanisms is the topic of the standardized *protocol specification*.

The global (N-1)-service provider can itself be an abstraction from two distributed (N-1)-protocol instances cooperating on top of an (N-2)-service provider. This descent stops at the interfaces to the physical transmission medium. Historically, this hierarchy of protocol layers was conceived and realized primarily in the other direction, i.e. from bottom to top.

Again, the local implementations involved need not really be structured according to the OSI architecture. In order to be conforming and interoperable, they merely have to behave externally as if they internally had this structure. Ultimately, the only thing that matters is the correctness of the bitstream they put "on the wire." This is one reason for the black-box principle in OSI testing.

The OSI Basic Reference Model identifies the seven layers shown in Fig. 1.2.

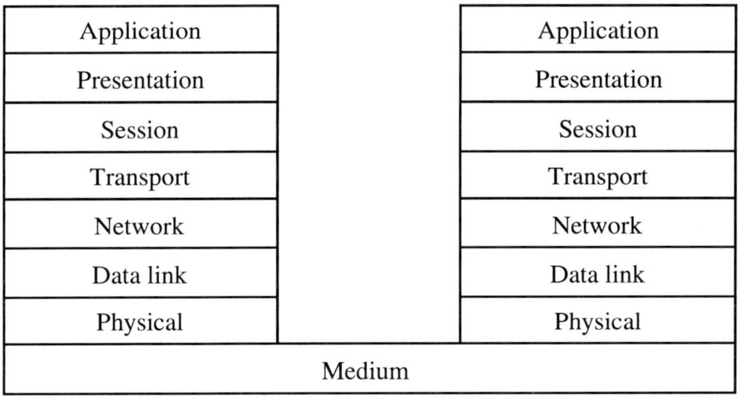

FIGURE 1.2. The seven layers of OSI

Their tasks can be outlined as follows:

The physical layer directly accesses the physical medium interconnecting the local systems. It is responsible for sending and receiving the bitstreams, i.e. for their physical transport over this medium.

The data link layer provides the undistorted data transport across a data link between adjacent nodes. This encompasses the detection (by means of added redundancy) and correction (by retransmission) of transmission errors.

From the third layer upwards, there is a duality of approaches, to wit a connection oriented approach and a connectionless approach, resulting in different OSI protocol standards for similar services. The connection mode seems to be able to claim more followers.

The network layer connects error-controlled data links along appropriately chosen paths to create end system connections. Some flow and error control are provided. Intermediate systems are only involved up to their third layer. The network layer permits the multiplexing of virtual circuits.

The transport layer provides additional flow and error control, and extends end system connections to connections between entities (e.g. application or session) within these systems. It offers a choice of five protocol classes differing in recovery, multiplexing and error correction abilities, but presenting one common upper surface: the transport service. The transport layer is the highest of the transport-oriented layers. The three layers above it offer application-oriented services beyond pure data transport.

The session layer provides primitives for dialog control. It permits communication to be structured into identifiable sections at different logical levels, which may extend beyond the duration of a transport connection. Moreover it allows for the dynamic control of sending, checkpointing and other capabilities.

The presentation layer offers facilities for the negotiation and use of transfer syntaxes for data exchange, which are independent from individual local data formats.

In the application layer, the actual cooperating and communicating top applications are supported by general, application-independent services, such as ACSE (association control service element) and CCR (commitment, concurrency and recovery).

The interested reader will find more details on OSI in general, and the OSI BRM in particular, in the standard [ISO 7498] and the recommendation CCITT X.200, as well as in the text books [HALS 92, JAAG 90, ROSE 90, STAL 90].

1.3.2 Application layer protocols, profiles

The OSI conformance testing methodology and framework (CTMF for short) presented in this volume can be applied not only to the OSI protocols in the narrower sense mentioned above, viz. the OSI layer 1-6 protocols and those providing the general application layer services, but also to specific application protocols that use this stack. Its concepts equally apply to a number of OSI-based and OSI-related application protocols, such as FTAM (file transfer, access and management), MOTIS (message oriented text interchange system), VT (Virtual Terminal), DTAM (Document Transfer, Access and Manipulation), TP (Transaction Processing), ISDN (Integrated Services Digital Network), MMS

(Manufacturing Message Service), GSM (Mobile Phone), ILL (Interlibrary Loan), etc. Test standards for these protocols, in accordance with OSI CTMF, are currently being prepared.

Three different factors have a combined and not necessarily very beneficial impact on standards. First, there is usually a vast spectrum of possible and mutually incompatible technical solutions in any problem area, which is, of course, the original reason for standardization efforts. Second, it is typical to avoid unnecessarily precluding implementation variants of a standardized product, as only the outward behavior counts. Last but not least, people working on standards want to minimize unpleasant clashes of interests.

Such conflicts are said to have even lead to "retaliatory compromises," which are not equally useful but rather equally unpleasant for all parties involved. At any rate, these factors frequently lead to protocol standards which provide innumerable parameters, functions, and function groups, including choices and combinations among these – the "catalog" approach. Still, the catalog approach seems to be preferable to the technique of simply remaining vague about debatable or technically demanding points. Of course, very few, if any, ambiguities are really introduced consciously.

Often the catalog approach offers more than users will ever need for their purposes. However, any implementor of the protocol would a priori be forced to include everything, both to pass conformance tests and to be able to successfully market a product. In addition to limited user needs, there is also a standards-inherent reason for a limitation of the choices and alternatives offered by a protocol: frequently implementations comprise a stack of adjacent protocol layers. In this situation, an (N)-protocol may happen to use only a true subset of the (N-1)-protocol's functions. The fact that users or next-higher protocols may only need and use a fraction of what a generous protocol and service offer is relatively harmless, when compared to the danger that two peer protocol entities may choose different selections from possible function groups and parameter values, which actually prevents them from interoperating.

To avoid such inefficiencies and incompatibilities, some useful combinations of the choices permitted by protocols have been predefined as, one might say, agreements on the use of standards. In order to avoid a user-specific multitude of incompatible solutions, such combinations, usually termed *profiles*, were soon adopted as actual standards: ISO issues International Standardized Profiles (ISP).

1.4 Standardization work on testing methodology

In this section we will take a look at two important international organizations involved in the standardization of protocol testing methodology and test suites, their roles, and their interrelations. The technical content of OSI conformance testing methodology will not be touched here, because it is the subject of the following chapter.

1.4.1 ISO/IEC JTC1

The *International Organization for Standardization* (ISO) was founded in 1946 for the advancement of industrial standards. Its members are national standardization bodies. They prepare proposals for standards and submit them for consideration by the other members and for international registration. Having obtained an ISO/IEC-number, the proposed standards pass through alternating phases of editing and voting, reaching the preliminary stages of Committee Drafts (CD) and Draft International Standards (DIS). Finally, if successful, they can obtain the status of an International Standard (IS). Usually, but not peremptorily, ISO member bodies adopt international standards as national standards of theirs. In 1988, the number of international standards passed by ISO or going through their preliminary stages exceeded 10000, elaborated by more than 3000 hierarchically structured committees.

Apart from the full members noted above, other organizations can become liaison organizations, with the right of speech but without vote; or they can become correspondent members with the status of an observer.

ISO and the *International Electrotechnical Commission* (IEC), founded in 1906 for the promotion of standardization in electric and electronic engineering, have overlapping fields of interest in the area of data processing. In order to avoid incompatible standards for the same appliances or procedures, ISO and IEC founded a first joint technical committee (JTC1) in 1987 for the field of information technology. Moreover, to avoid new overlappings and to cope with existing ones, close contact is maintained between JTC1 and the International Telecommunication Union (ITU).

In ISO, the ISO/IEC JTC1 took the place of the former Technical Committee TC 97 (Information Processing Systems, one of about 200 in ISO), while in IEC it replaced the TCs 47B and 83. As previously in TC 97, its standardization areas are currently divided among sixteen subcommittees. Of particular interest in the context of communication protocols are SC 6 (Telecommunications and Information Exchange between Systems) and SC 21 (Information Retrieval, Transfer and Management for OSI). At an intermediate level, above the SCs, JTC1 is structured into 4 sections, of which section "Systems" comprises SC 6 and SC 21. Subcommittees themselves are usually structured into working groups (WG). Each full member of ISO can delegate a representative to each TC or SC or WG, either as an active voting member (P-member) or only as an O-member, i.e. one who is merely kept informed.

In addition to the four sections mentioned above, JTC1 comprises among others a Special Group on Functional Standardization (SGFS), which deals with protocol profiles and profile testing. They produced the technical report [ISO 10000]: International Standardized Profiles, comprising the parts 1: Taxonomy Framework, and 2: Taxonomy of Profiles.

THE DEVELOPMENT OF IS 9646

The first work within ISO on the topic of "Conformance Testing Methodology and Framework" began in 1984. As pragmatic research and development work had been tackled as early as the late seventies, a considerable amount of technical knowledge and experience was available. From these beginnings, the methodological framework for protocol conformance assessment has been conceived, consolidated and perpetuated in Project 1.21.23 of JTC1. This work was done in cooperation with CCITT, for the most part in common meetings.

At the same time, work was started on test suites for existing protocols based on this methodology. X.25 and MHS were among the first protocols considered for these activities. For the methodology developers, this provided early feedback on the usefulness and acceptance of their emerging standard, in particular with respect to the test notation. For the test suite writers, this meant that the test notation they were using changed more frequently and more drastically than they appreciated – a consequence of both feedback-initiated and general improvements, extensions, and simplifications.

The relatively long instability of the test notation TTCN had three major causes:

For one thing, TTCN is a very complex language – more complex than some programming languages. Development of a sufficiently stable programming language requires a considerable amount of manpower, in particular if it is developed by a relatively large body, and if, in addition, it is subject to and influenced by user wishes during its evolutionary stage. Considering, too, that the developers of the language could approach these efforts only as a side job, over the course of a few weeks per year, the work by necessity had to spread over some years.

Next, TTCN, of all parts of [ISO 9646], is the most formal one, and the decisive one for the actual assignment of test verdicts. While slight inaccuracies in general descriptions of concepts are often easily accepted (or even desired, for reasons of compromise), inaccuracies in the test notation can lead directly to erroneous test results.

Finally, there is a specific historical cause for the slow stabilization of TTCN: the notation was originally conceived only as an abstract and informal language, to be complemented later by a formal and binding executable test description language. Gradually this dual approach was abandoned in favor of TTCN as the single test notation.

Meanwhile, several test standards, parts of standards, and addenda, based on [ISO 9646], have emerged or are being prepared for concrete protocols; in particular, on test suite structure & test purposes (TSS&TP) and on PICS proformas, e.g. ISO/IEC 8073 AD3: Transport PICS and ISO/IEC 10025/2: Transport TSS&TP. The preparation of final versions of standardized test suites for protocols could not be started on before TTCN reached the status of an international standard, which was the case in 1992.

A more recent standardization effort is Project P.1.21.54 of JTC 1/ SC 21, which in conjunction with ITU-TS (Question Q.10/X), pursues the aim of applying formal methods to protocol conformance testing.

1.4.2 ITU-T (formerly CCITT)

CCITT, the Comité Consultatif International pour Télégraphique et Téléphonique, was one of three permanent committees of the ITU, the International Telecommunication Union. In 1993, it was renamed ITU-T, the Telecommunication Standardization Sector of ITU. The CCITT recommendations, as they were previously called, will now be named ITU-T recommendations. They comprise technical standards, tariffs, administration and terminology regulations. In spite of their officially non-obligatory nature they are factually automatically adopted as binding regulations by the postal and telecommunication authorities (or companies) involved.

Originally, apart from the irregularly issued provisional recommendations, the entire set of the latest CCITT recommendations were passed by the plenary assembly at four year intervals, each time under a different "color code." The 1984 recommendations form the Red Book, those of 1988 the Blue Book, and those of 1992 the White Book. From 1993 on, new and revised recommendations can be approved at any time.

The technical work of ITU-TS is performed in Study Groups (SG), which are subdivided into Working Parties (WP). The standardization of protocol tests and their methodology is dealt with in the SGs 7 (Data Network and Open System Communications), 8 (Terminal Equipment and Protocols for Telematic Services), 10 (Languages for Telecommunication Applications), and 13 (General Network Aspects).

Recommendations X.290 to X.294 of the White Book (OSI Protocol Conformance Testing Methodology and Framework for Protocol Recommendations for CCITT Applications) were published in accordance with the parts 1 to 5 of [ISO 9646]. The ITU-T recommendations corresponding to the second versions of parts 1 through 5 as well as to parts 6 and 7 of the ISO/IEC standard 9646 will be published in 1994 and 1995 as X.290 through X.296, respectively.

1.5 Interoperability testing

Interoperability testing has occasionally been recommended as an alternative to, or even a preferable substitute for, conformance testing, sometimes in quite controversial discussions. Over time, the original view of the two methods as competitors has given way to a view in which they are considered to be complementary measures which increase confidence in both the "correctness" and the interoperability of products. Due to its importance, be it as a competing or as a companion approach, we will try to provide our readers with an introduction to interoperability testing.

1.5.1 What is interoperability testing?

Currently, there does not seem to exist any generally accepted, formal or semi-formal definition of interoperability testing (which we will henceforth refer to as IOT). However, the authors expect that eventually an "ITMF" standard will emerge analogous to the CTMF standard ISO 9646.

Let us, by way of substitution, list some possible views of interoperability and some types of activities which were termed IOT. *Interoperability* can be interpreted as:

- what users expect from the product they purchase, which may be more than what is demanded by the standards, and which may include aspects excluded from conformance testing, such as performance, robustness, and reliability;
- proper functioning in the system where the product is finally installed;
- compliance with a global service specification, which may include more than the rather locally oriented description typically provided in service definition standards, and which may possibly include "real effects," for example, on a data base accessed by means of the product;
- interoperability of applications which use the product, i.e. an "embedded" view of interoperability.

Interoperability still seems to be a very informal and practical concept, one that is intuitively clear, but that few have yet tried to formalize. This may, however, change rapidly, as soon as some "ITMF" activity brings enough tangible structure and concreteness into this field.

The term IOT has already been applied to activities

- in ad-hoc vendor demonstrations,
- in permanent groups established for common demonstrations and registration,
- in commercially offered IOT services.

The purpose of these activities ranges from the creation of registers or databases (who cooperated how well with whom?) via the formation of informal cooperation groups (all our members were, are, or try to be able to interoperate), to services involving actual commitments (if the registered products of our members/clients fail to interoperate, we will determine the problem and have the bug fixed, or some other appropriate action).

1.5.2 Potential IOT problems

When considering the vague circumscriptions given in the previous subsection, some potential drawbacks of IOT rapidly emerge. Some of these can be viewed as disadvantages when compared to conformance testing. They will be better understood after Chapters 2, 3, and 4 have been studied.

As far as interoperability in terms of the eventual productive application of the implementation is concerned, an IOT environment may not simulate the target system and environment well enough.

If the possibilities for control and observation are less than in conformance testing, then IOT may boil down to some passive monitoring with little room for purposeful investigations.

Interoperability not being transitive, the number of necessary interoperabilty tests grows with the square of the size of the "population" of potentially compatible implementations.

As IOT takes place with peer entities which are ordinary implementations, there may not be any way to produce errors intentionally. This makes it impossible to test responses to invalid PDUs.

1.5.3 The pros of IOT

It is possible that an implementation which underwent thorough conformance testing is not able to interoperate with a wide number of products on the market. This may be caused by reasons other than errors in the implementation or general deficiencies of conformance testing, such as the following:

- The existing products may have the same misunderstanding of some important point of the protocol, perhaps as a result of harmonization.

- Similarly, the standard may be ambiguous, and the new implementation may have chosen a new interpretation of the point in question.
- The coverage of the applied test suite may be to small.

At any rate, IOT is capable of revealing this problem early, although it may not provide enough information to decide what action should be taken to resolve the problem. Resolution testing and/or some discussion about, or improvement of, the standard may be required.

IOT may be cheaper than conformance testing, as long as the population of different implementations tested against one another is not too large.

IOT may be appropriate to test for the fulfillment of user expectations not covered by the protocol standard, or otherwise excluded from conformance testing.

An IOT environment may simulate the target system and environment more closely than a conformance testing environment.

Even with only the above, admittedly quite vague discussion, interoperability testing seems to be a natural second step to be taken after conformance testing. It will be even more promising, once a comparably rigorous philosophy, methodology and notation, as exist for conformance testing, have been developed.

Considering the powerful and very specific facilities to express control and observation in arbitrarily distributed environments which TTCN provides, it seems that this notation could also be used to specify interoperability tests.

CHAPTER 2. THE OSI CONFORMANCE TESTING METHODOLOGY

2.1 OSI conformance testing projects

Presently, three main approaches prevail in the field of ISO conformance testing, which are consequently treated in three different projects within ISO/ IEC JTC 1/ SC 21/WG 1:

- In Project 1.21.23, the oldest testing project, a heuristic OSI conformance testing methodology including a test notation was developed. This work began in 1984 and resulted in a multi-part international standard [ISO 9646]. We will often refer to this standard as 9646 or *CTMF,* for *conformance testing methodology and framework.* It can be considered as finished, since future major extensions will only be developed under other ISO testing projects. Of course, the clarification and definition of remaining ambiguities in 9646 will be a continuing effort.
- Project 1.21.54, launched in 1990, pursues mathematically based formal approaches to OSI testing. The main document [FMCT 94] developed thus far is a working document entitled *Formal Methods in Conformance Testing* (FMCT), which is intended for protocols specified by means of formal description techniques (FDTs). FMCT is not being developed as a replacement for CTMF. However, there are many points of contact, where the FMCT document either supports 9646 or dissents from it. The main work on FMCT includes the definition of a test methodology with formal descriptions of test entities and the automatic derivation and validation of test suites from formally specified protocol entities.
- In a new work item, planned to start in 1994 or later, an advanced *open systems assessment methodology* will be developed, which will include the assessment of open systems claiming conformance to open distributed processing (ODP), open systems environment (OSE) profiles and network management standards. As a starting point, the project uses the terminology and basic concepts provided by CTMF. However, in addition to CTMF, the new project also deals with different types of interfaces, such as application programming interfaces, man-machine interfaces, storage interfaces and robotic interfaces, which cannot be treated by CTMF. Furthermore, the open systems assessment methodology will cover its relationship to passive testing, performance testing and interoperability testing, which are also not part of CTMF.

We will present and discuss neither the open systems assessment methodology (more information about this topic is given in the study report [PROST 93]), nor formal methods.

The present chapter will be devoted to a presentation of the basic concepts and ideas of the conformance testing approach developed in CTMF. CTMF consists of the following seven parts, for some of which addenda or a second edition already exist:

- Part 1: General concepts [ISO 9646-1],
- Part 2: Abstract test suite specification [ISO 9646-2],
- Part 3: The tree and tabular combined notation [ISO 9646-3],
- Part 4: Test realization [ISO 9646-4],
- Part 5: Requirements on test laboratories and clients for the conformance assessment process [ISO 9646-5],
- Part 6: Protocol profile test specification [ISO 9646-6], and
- Part 7: Implementation conformance statements [ISO 9646-7].

For didactic reasons, we will not present the material exactly in the order suggested by these parts. Rather, we will organize their contents according to common, central topics, while at the same time avoiding some of the repetitions found in the standard.

2.2 The field of the OSI conformance testing methodology

One of the main prerequisites for the realization of the aims of OSI is the availability of systems tested and recognized to be conforming to protocol and profile standards. The world-wide comparability and acceptance of the test results produced by various test laboratories is important both for vendors and procurers of protocol implementations. It is only by means of effective and well-accepted schemes of testing and certification that costly repetitions or variations of conformance tests of the same implementation can be avoided.

Two concrete measures are being taken to achieve this aim: standardized *abstract test suites* (ATSs) and a standardized testing methodology.

ATSs are being developed, written in a standardized test notation. They are defined and applied within the framework of a standardized testing methodology, describing the appropriate test methods and testing procedures. We will see that along with the ATS standards, the testing methodology calls for additional protocol-specific standards. The definition of the test suite structure and the test purposes guides the ATS production. Certain proformas permit the ATS to be adapted, in a well-defined way, to the parameters chosen and options realized in the implementation submitted for test. All these topics will be treated in detail in this chapter.

A number of important aspects of protocol testing have not been taken into account in 9646, either inadvertently or deliberately. These include:

- interoperability testing,
- passive testing,
- performance,
- robustness,
- the realization of service primitives,
- internal aspects of the system implementation,
- the system environment,
- procurement procedures,
- test certificates,
- accreditation of test labs,
- testing tools, etc.

Some of these excluded aspects will be treated in the open systems assessment methodology. Conformance testing will always be limited, for technical and theoretical as well as for economic reasons. Generally, no guarantee can be given that an implementation absolutely conforms to the relevant protocol standards. However, conformance testing with positive test results will increase the level of confidence in a protocol implementation.

The notion of *conformance* in OSI refers to protocol standards, not to service standards, because only the former are considered to define externally visible valid behavior. Nevertheless, work is progressing on service interface standards. However, this has not yet resulted in any finished standard.

The black-box philosophy of the OSI Basic Reference Model is reflected in the statement that "only the external behavior of Open Systems is retained as the standard of behavior of real Open Systems [ISO 7498]."

In this respect the protocol testing sector is really in a quandary, in so far as the closer the testers want to look at the implementation's behavior, the more they demand access which is not required by OSI standards to be available. In doing so, they contradict the liberty of implementation granted in the protocol standards. As an additional complication, whenever a stack of protocol entities from several adjacent layers is implemented in an integrated product, access to the lower layers is particularly limited.

The closest way to examine an implementation "from the outside" is by having the full control of both the upper and lower boundaries, often realized as service interfaces. This configuration was deemed to be so improbable for conformance testing, except perhaps in first-party testing, that it is not considered in 9646. The way taken out of the dilemma was the definition of a choice of several test methods with various degrees of access to the implementation, and with varying requirements and constraints placed on the implementer. The fewer the requirements for automatic manipulation that are given, the more manual work is required from the test operators during the execution of tests. Clients can choose the test methods to which they want to submit their implementation, assuming that an ATS exists for all their test methods.

OSI conformance testing methodology is concerned with the various phases and activities related to the conformance testing process, comprising

- the development of proformas for implementation conformance statements and implementation extra information for testing,
- the specification of abstract test suites for OSI protocols,
- the adaptation of proformas and ATSs to match profile requirements,
- the realization of test tools and means of testing, capable of executing particular test suites,
- the conformance assessment process, performed at a test laboratory for a client, with a protocol conformance test report as the ultimate result.

Part 1 of [ISO 9646] deals with all activities. It gives a general overview, providing definitions and descriptions of important notions and concepts of OSI testing. These will be treated in Section 2.3.

Part 2 states requirements about the specification of abstract test suites. It gives guidance on their composition and on the choice of the proper test method. These considerations are independent of the test notation. We will treat these topics in Sections 2.5 and 2.6.

Part 3 describes the recommended test notation TTCN, which we will present in detail in Chapters 3 and 4.

Part 4 deals with the requirements placed on test realization, i.e. on the production of a means of testing. We will consider test realization in Section 2.7.

Part 5 describes the requirements related to conducting the conformance assessment process, in particular the documents exchanged between the test laboratory and the client, and the steps leading through the test campaign to the test report. We deal with these in Section 2.8.

Part 6 presents concepts of the OSI protocol profile conformance testing methodology, aiming in particular at the economical specification of profile test suites. Sections 2.3 and 2.4 provide information on this subject.

Part 7 specifies requirements with respect to the design and use of implementation conformance statement proformas. The concept of implementation conformance statements and their related proformas or questionnaires are treated in Sections 2.3 and 2.4.

The OSI protocol testing methodology has been extended by a protocol profile testing methodology which is gaining more and more in significance, due to the growing number of profiles for which test suites have to be provided. The fundamentals of the profile concept were developed by the special group on functional standardization (SGFS) of ISO/IEC JTC 1, and can be read in their technical report TR 10000 [ISO 10000-1, ISO 10000-2].

Protocol profile testing methodology was textually incorporated in the form of amendments to the existing parts of 9646. In addition, it was published as a separate part 6 entitled "Protocol Profile Test Specification." In this section we confine ourselves to a short description of the essential points of the profile testing methodology.

In general, profile testing does not require separate test purposes, test cases, test methods and test suites. Instead, all the concepts and documents developed for testing protocol implementations are basically also used in profile testing. More is needed when the profile places additional requirements on the protocol specifications or when it defines interrelationships between protocol layers that are not covered by the related protocol specifications alone.

The reader of 9646 who is interested in profile testing may get a little bit confused since the text on profile testing is distributed over different parts of 9646 and contains many new terms which have partially overlapping meanings. For this reason we have reorganized the material on profile testing, so as to present it in this section, in conjunction with information on protocol testing.

2.3 General concepts

2.3.1 *Conformance requirements*

In the OSI sense, real systems are called *conforming* if their externally visible behavior fulfils the *conformance requirements* stated in the relevant OSI, e.g. protocol, specifications. Conformance requirements hence form the cornerstone of the entire testing methodology. They define the demands that can be imposed on implementations. The clauses in a standard defining conformance usually list the technical parts which directly and normatively define the protocol behavior. This excludes parts like the introduction, scope considerations, references, abbreviations, definitions, notational conventions, informative annexes, etc. One should not confuse the conformance requirements, the subject matter of the technical clauses of a protocol standard, with the *conformance clauses*, which merely point to these technical clauses.

Various types of conformance requirements are distinguished from one another:

- *mandatory requirements*, which always have to be observed,
- *conditional requirements*, which have to be observed under conditions defined in the standard (frequently a dependence on previous choices of parameter values and options or a prescription or interdiction of certain combinations of choices),
- *optional requirements* or shorter options, which can be freely selected, but possibly subject to restrictions explicitly given in the standard.

This distinction is far from being precise, because it allows the standard to label conditional requirements optional. Suppose the standard confronts the implementer with the strict alternative of implementing either a feature A or a feature B. One could call them optional, with the additional requirement to choose at least one but not both. With equal justification one could e.g. call A optional without any requirement imposed, and B conditional with the condition prohibited if A was selected, mandatory if not, etc.

More important for the conformance testing process is the distinction made between *static* and *dynamic conformance requirements.* Static conformance requirements specify the permitted combinations of capabilities implemented in a real system. These often include a minimal set of capabilities necessary for interworking. Static conformance requirements come at various levels of detail. They may deal with broader features such as protocol classes or functional units, but they are also concerned with details such as value ranges for size parameters or timers. They can be single-protocol requirements as well as multi-protocol dependencies, the latter appearing mainly in upper layer protocol standards.

Dynamic conformance requirements usually form the largest part of OSI specifications. They specify the observable (i.e. communicative) behavior of the implementation permitted by the standard, thus defining the "actual protocol," i.e. valid sequences of PDU exchanges, timing, structure and format of PDUs. This specification is usually structured according to major protocol or service aspects, such as protocol phases, handling of normal situations vs. error handling, functional units etc.

By definition a *profile* is based on its related *base specifications,* on which it will usually place additional constraints. Thus, a profile implementation's behavior becomes a true subset of the behavior permitted by the base specifications.

Conformance requirements play the same important role in testing a profile as in testing base specifications. Therefore, the same concepts for static and dynamic conformance requirements are used, and conformance requirements are expressed as far as possible by references to conformance requirements in base specifications. Depending on the profile specific restrictions of the base specifications, the status of referenced conformance requirements may change in a profile as follows:

- mandatory remains mandatory, i.e. it cannot change;
- conditional may remain conditional or may change to mandatory, optional, or excluded;
- optional may change to mandatory or conditional, or one of the following profile specific attributes:
  - out of scope, which classifies options that are not used by the profile;
  - excluded, which classifies options that are prohibited by the profile;
- not applicable remains not applicable.

Of course the reservations about the classification of protocol conformance requirements mentioned before hold equally for this classification, with *out of scope* as another candidate for ambiguity.

2.3.2 *Implementation conformance statements*

For the purpose of conformance testing, conformance requirements are presented in the form of questionnaires in several proformas, which have to be completed by the client to become an *implementation conformance statement* (ICS). An ICS is a statement of the capabilities realized in a particular implementation and the parameter values chosen for it.

The client produces an ICS by filling out an *ICS proforma* (see Section 2.4). If the data given in the ICS are consistent with the static conformance requirements of the protocol specification, the test laboratory can use them to test the IUT against the appropriate dynamic requirements. In practice, this requires the selection of the applicable test cases and their instantiation with the right parameter values. Inappropriate tests, e.g. those concerned with options not implemented, are thus deselected before testing begins.

If an IUT realizes a single protocol, then a *protocol implementation conformance statement* (PICS) is required for testing.

If an IUT contains a set of interrelated protocols, e.g. a stack of protocols from adjacent OSI layers, then a *profile implementation conformance statement* (profile ICS) is needed. A *profile ICS* includes a PICS for each of these protocols, possibly ICSs for relevant information objects, a profile specific ICS and finally the profile requirements list (RL).

The *profile specific ICS* provides additional conformance statements derived from the profile specification that are not given in the other ICSs. The *profile RL* defines further restrictions on the PICSs and *information object ICSs*, which the client has to consider when completing the proformas. The style and contents of profile RLs are not completely regulated. However, it is recommended that any duplication of the PICS proforma be avoided, and that all profile restrictions be handled by the profile RL. One possible form of a profile RL is simply a list of the additional constraints which affect the allowed answers in the PICS proforma tables. A more appropriate and efficient form however is a modification of only those PICS proforma tables which are affected by the profile restrictions. In Figure 2.1 the relationship between these different documents, omitting information object documents, is illustrated.

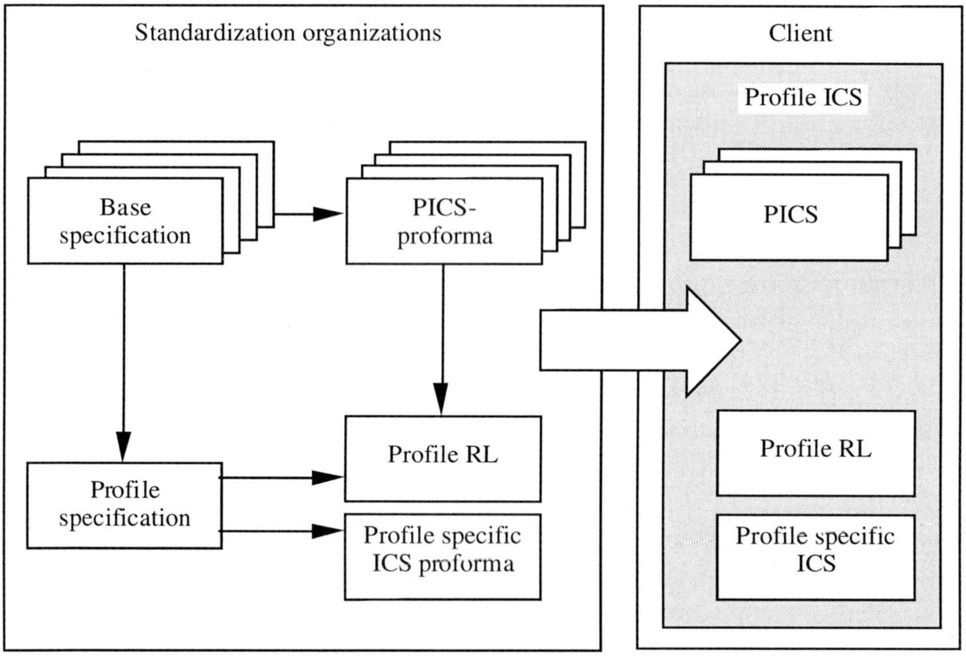

FIGURE 2.1. Profile ICS documents

PICS proformas are expected to be developed by the groups having defined, or defining, the respective protocols. They appear either as a separate part or a normative annex of the respective protocol standard. For PICS proformas, the ISO or CCITT copyright is usually waived, so as to permit unrestricted copying and use.

2.3.3 Extra information for testing

In order to perform conformance tests, test laboratories need some further information beyond the specification related information collected in the ICS. This is referred to as *implementation extra information for testing* (IXIT), which is an *IXIT proforma* filled out by the test laboratory and the client. IXIT proformas are produced by test suite specifiers, test tool developers and the test laboratory for each abstract test suite to be applied. Analogous to the different ICS types, *Protocol IXIT* (PIXIT) *profile IXIT* and *information object IXIT* are distinguished. Of course there should not be any inconsistency, nor any duplication, between ICS and IXIT documents.

IXITs are used to specifiy additional testing information related to an OSI specification that cannot be derived from the specification itself. They typically contain information such as SUT and tester SAP addresses, and purely test-specific timer or parameter values, which are not mandated by the protocol standard.

Profile IXITs are needed for collecting additional testing information of profile IUTs. A profile IXIT includes a PIXIT for each of the protocols of the profile, plus IXITs for relevant information objects, a *profile specific IXIT* and finally the *profile IXIT requirements list* (XRL). The profile specific IXIT provides additional testing information which is not contained in the other IXITs. The profile XRL defines further restrictions on the PIXITs and information object IXITs, which the client has to consider when completing the proformas.

The client is additionally required to prepare a *system conformance statement* (SCS), listing references to specifications of the implemented profiles, protocols and information objects, to related profile specific ICSs, PICSs and to related test reports. The SCS also provides information about the configurability of the system under test.

These notions given, we can now provide the definition of conformance of an IUT: An implementation is said to be *conforming* if it satisfies the static and dynamic requirements of the relevant standards referenced in the SCS, in consideration of the capabilities stated as implemented in the ICSs.

2.3.4 Types of testing

In CTMF, four types of conformance tests are defined, which differ from one another in their purpose and level of detail:

- basic interconnection tests (BIT),
- capability tests,
- behavior tests, and
- conformance resolution tests.

Basic Interconnection Tests

Basic interconnection tests (BIT) have a very limited scope. They are used to determine whether interconnection is possible at all. They may be performed as an initial stage of conformance testing, in order to find out whether further in-depth testing is reasonable. Their completion or their failure can constitute a *negotiated exit point* (NE) of the test campaign.

Basic interconnection tests

- demonstrate the IUT's ability of connection set-up, including addressing and possibly negotiation, and
- provide a filter function for further tests, permitting a negotiated exit if BITs are failed.

An abstract test suite may contain a list of test cases to be selected as BITs if this is agreed upon between the test laboratory and the client. In the test suite, BITs are optional and need not exist in the form of an explicit test group.

Capability Tests

Capability tests form a subset of the test cases of an ATS. They are also optional and may constitute a separate test group. As a second preliminary stage of conformance testing, they have the limited purpose of validating that all the capabilities claimed in the PICS to be implemented can be observed. Capability tests, too, can define a negotiated exit point of the test campaign. They provide

- validation of the IUT's capabilities against the static conformance requirements and the PICS;
- in conjunction with behavior tests, the basis for a claim of conformance;
- a filter function for in-depth behavior tests, permitting a negotiated exit if capability tests are failed.

Behavior Tests

Behavior tests form the main part of an abstract test suite. They test an implementation as thoroughly as technical and economic conditions permit. Behavior tests are expected to cover the entire spectrum of the dynamic conformance requirements specified in the relevant protocol or transfer syntax standard(s). This includes the validation of the behavior of the IUT in response to invalid protocol behavior intentionally exhibited by the tester.

CONFORMANCE RESOLUTION TESTS

Conformance resolution tests are special tests for purposes going beyond ordinary behavior testing, such as

- investigating particular aspects of the IUT, e.g. to locate causes for invalid behavior,
- testing with non-standardized test architectures,
- resolving problems encountered in ordinary ATS testing,
- exploiting special facilities of the system under test, in particular for controlling SUT-internal states or events. This may concern the creation of resets, backpressure, response delays or busy states.

Although defined in 9646, they are then explicitly declared as out of the scope of the standard, and are not considered any further.

2.3.5 *Test results*

In this section we distinguish three different types of test results according to their granularity and scope. These test results may be positive or undesired in various respects (e.g. disclosing faults in the ATS or the IUT). We analyze the requirements which are usually imposed on these test results.

TYPES OF TEST RESULTS

A *test outcome* is comparable to a complete trace of a test case execution, i.e. the sequence of events that occur and are recorded by the tester in this process. It includes in particular the entire input to and output from the tester at the access points for testing, called the *points of control and observation* (PCO, see 2.5.3). A test case should be able to cope with all possible test outcomes, even in case of serious flaws in the IUT. It does so by means of event patterns capable of matching even unexpected input from the IUT. The test outcomes covered by the test case are called *foreseen*.

An *unforeseen* test outcome constitutes a *test case error*. It should be recorded in the conformance log and should lead to an *abnormal termination* of the test case. Note that what is then recognized directly is an error in the executable test case (*executable test case error*), which may originate from an error in the specification of the abstract test case (*abstract test case error*), or from an error in the derivation or execution of the executable test case.

Abnormal termination is recorded not only if it occurs after a test case error but also for any other reason, such as network problems, tester system problems, etc., which cause the test operator or MOT to stop execution before the test case has completed.

The ATS assigns to each individual and foreseen test outcome one *test verdict*. The possible test verdicts are *pass*, *fail* and *inconclusive*, and their meanings are as follows:

- Pass means that the observed test outcome shows that the test purpose has been fulfilled and that the IUT has displayed only valid behavior as specified in the relevant protocol standards. It gives evidence of conformance to the requirements.
- Fail means that the test outcome proves that some conformance requirement has been violated, either the one given by the test purpose or some other.
- Inconclusive means that neither pass nor fail can be assigned to the test outcome, because the test purpose has not been fulfilled, yet no protocol errors have occurred. This typically happens when the underlying provider signals problems (e.g. provider Disconnect or Reset primitives).

Note that discussions continue about what actually can be demanded in a test purpose when dealing with a non-deterministic point in the protocol specification. Is it, for example, a fail, a pass, or an inconclusive if the IUT stubbornly declines a connection establishment request, e.g. for reasons of local resource shortage, and if the protocol lists this as a valid alternative? Are there alternatives which can reasonably be demanded (e.g. acceptance of connection) and others which cannot (e.g. signalling an irregular local condition)? This topic is addressed in more detail in Subsection 4.6.3.

The observed test outcomes and the assigned verdicts of all test cases executed, as well as any other observations deemed worth noting, have to be recorded by the test laboratory in a *conformance log*.

The final step in the conformance assessment process is the documentation of the test results in the *system conformance test report* (SCTR) and one or more *protocol conformance test reports* (PCTR). A PCTR is produced for each protocol realized in the IUT and tested in the test campaign(s). It contains the results of each test case and gives references to the conformance log and all other relevant test documents. The SCTR is a summary of all verdicts assigned, rating the achievement of conformance of the system under test. In Subsection 2.8.4 we will present details about the structure and contents of these reports.

General Requirements on Test Results

CTMF attaches great importance to the stability of test results: they are expected to be *repeatable*, *comparable* and *auditable*, at least to a reasonable degree. Otherwise the efforts invested to make test results internationally accepted will be doomed to fail from the outset.

Of course, it is desirable that all executions of the same test case on the same IUT give identical results. But time variations in real executions and non-determinisms in protocol and service specifications can make it impossible to write test suites which produce identical results in each repetition with the same IUT. Nevertheless, test case writers, test tool implementers and test laboratory staff are required to minimize the chances of variations in the test results. This holds in particular with respect to test verdicts, which should be identical whenever the underlying provider and the IUT exhibit in principle the same behavior. However, it is not formally defined in CTMF when behaviors should be called the same or equivalent. FMCT will certainly undertake efforts in this direction, cf. Section 2.1.

The question of the comparability of test results comes into play whenever test reports can be issued by several test laboratories, whether these are run by the supplier, the user or a third party. Important factors improving the comparability of test results are careful design, unambiguous specification, and faithful implementation of test cases, means of testing, test laboratory procedures, and test report production. Regardless of the desire to leave a useful degree of flexibility in the realization of these documents, systems and activities, the requirements stated by the protocol and CTMF standards should always be respected, if test results are to remain comparable.

Since financial interests usually will be involved in the conformance assessment process, there will occasionally arise the need of reviewing the results of a test campaign. The auditability of test results is warranted by the orderly keeping of a conformance log, from which all inputs, outputs and other test events can be read. This permits the review of all observed test outcomes and makes it possible to verify the correct assignment of all test verdicts.

2.4 Testing documents

2.4.1 The ICS proforma

In this subsection, we will give more details on the ICSs and ICS proformas. In particular, we will deal with protocol capabilities and how they are reflected in the structure and contents of the PICS proforma.

CAPABILITIES

A protocol capability is a set of functions which, if implemented, are implemented together. Capabilities also are referred to as functional groups, protocol classes, etc. A conforming implementation may or must support them, depending on their status. In OSI specifications, each capability is assigned a status, which can be any one of the following:

- *mandatory*, if the OSI specification requires that the capability is realized in every implementation.
- *optional*, if the capability may be implemented or not. This choice may be subject to conditions, i.e. here again we have an overlap between optional and conditional. An informative annex in [ISO 9646-1] explains that
  - *Boolean* options are the truly optional capabilities; they may be implemented or not, the choice being independent from any other choices;
  - *mutually exclusive* options mean a choice of exactly one out of a set of possibilities, while
  - *selectable* options mean a choice of m (≥1) out of n (≥m, ≥2) possibilities.
- *prohibited* or *excluded*, which does not refer to a capability as a whole (because then the capability would not be described in the standard in the first place), but rather to a situation arising dynamically where the capability shall not be applied. Note that in profiles, some capabilities can be excluded, too.
- *out-of-scope*, which indicates that the capability is outside the scope of a profile and hence irrelevant for testing.
- *not applicable*, if no requirement can be expressed in a given context – a status value which permits many interpretations.
- *conditional*, if the requirement on the capability depends on the selection of other optional or conditional capabilities. The uniqueness lies in the condition being more complex than the categories "selectable" or "mutually exclusive" mentioned above. Conditional capabilities may be associated with predicates or conditional expressions, the evaluation of which will result in a more definite status, preferably mandatory, truly optional, or prohibited.

It is not accidental that 9646 defines types of conformance requirements similarly to status values of capabilities, because the necessity or choice of implementing a capability is a conformance requirement. It might be worthwhile to create a common and more formal approach to these notions in the future. Standards users have been reported, at times, to be confused by these notions, especially when it comes to the definition and implementation of profiles, where the interdependencies are often even more involved.

Technically, capabilities can be grouped in the following categories:

- major capabilities,
- supported PDUs,
- supported fields within PDUs,
- supported timers,
- encoding variations,
- negotiation mechanisms,
- error procedures.

Major capabilities are larger units of functionality into which the OSI specification is structured, e.g. protocol or service classes, service elements, functional units. They may correspond to certain groups of PDU types or to some aspects of several PDU types, but they need not. Initiator-responder capabilities are the abilities to initiate communication or to respond to such initiatives. They differ in asymmetric protocols. Timers and protocol parameters are capabilities usually of the mutually exclusive option type. Their names and permitted ranges are defined in the protocol specification. The support of the various PDU types may be related to the selection of major capabilities and may differ with respect to transmission or receipt, as with initiator-responder capabilities. The support of PDU fields addresses in detail the ranges and types supported in either direction. If there is no implementation latitude at all, the support value is simply either yes or no. Encoding variations may require separate consideration. Other capabilities are those to negotiate particular options, and manners of protocol error handling.

General Issues

The ICS proforma reflects the range of choices left to the implementor by the protocol, profile or information object specification, i.e. it lists the implementation options with respect to capabilities as well as the legitimate parameter value ranges, from which the implementor can choose. For the sake of completeness, however, the mandatory capabilities are also listed.

ICS proformas are to be used by implementors or suppliers of implementations primarily, because they will document the implementation's characteristics. Options or values which can be chosen after acquiring the IUT may be exceptions to this rule. Test suite writers may use the ICS proforma as a compact overview of the protocol options and value ranges, to prevent unduly restricting the latitude permitted of implementations. Profile specifiers may make similar use of ICS proformas. Even customers in need of an overview of the possible features of a product may use them. The completed ICS proforma, i.e. the ICS, will be used in the conformance assessment process as follows:

- in the static conformance review,
- in the selection of test cases and the definition of their parameter values,
- and finally as a reference document in the test results analysis.

An ICS proforma is a list of questions about the realization or support of the capabilities in the implementation, plus fields to contain the answers. The supplier of the IUT will indicate for each capability whether it is implemented or not. In some cases more specific answers are permitted, such as the implementation of a well-defined part of the capability (e.g. only implemented as a responder). The answers to these questions will be submitted to the static conformance review. There is a very close relationship between static conformance requirements in the OSI specification and the items in the ICS proforma. In principle, for each testable conformance requirement there exists a corresponding ICS question, which references the corresponding clauses in the standard.

Of course, an IUT is not conforming if a mandatory capability is not implemented. But it is provided that in an ICS it can be explained why it is not implemented. One reason may be that from existing defect reports on the OSI specification it can be inferred that the status of the capability shall be changed to prohibited in the given context. It is conceivable that even some kind of conformance certificate might be obtainable under such circumstances. But this is beyond CTMF.

Even if a capability is optional, the OSI specification may permit situations where an IUT not supporting it is confronted with a PDU related to this capability. This could be the case if the PDU is coming from another implementation which does support this capability. The protocol has to specify one or several permitted reactions in such a case, and the ICS may ask about the choice of reaction realized in the IUT. The ICS proforma must indicate the possible choices and the interdependencies between the different choices, if necessary in separate tables indicating clearly the predicate variables and related ICS entries.

THE STRUCTURE OF THE ICS PROFORMAS

The structure of an ICS proforma is shown in Figure 2.2. The first section of the ICS proforma contains an *ICS corrigenda table*, which may be used by the proforma supplier to add the corrections applying to the last ICS proforma version until publication of a new version. The standard does not explicitly tell whether the corrections are specified or only referenced in this section, but we take the second interpretation. The second section gives instructions for the client regarding the completion of the ICS proforma, and provides information about its structure and purpose, as well as about the symbols, abbreviations and terms used in the proforma.

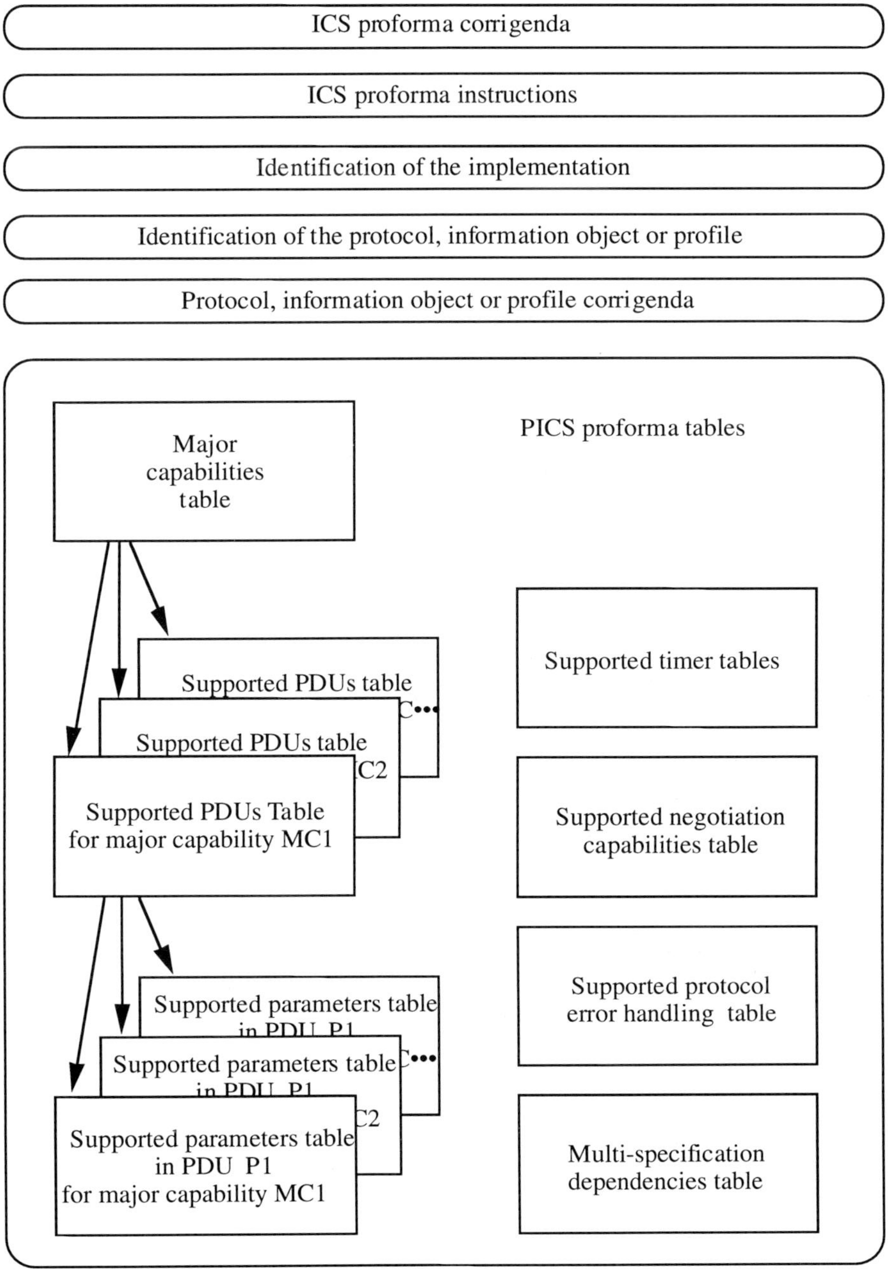

FIGURE 2.2. Structure of an ICS proforma

The third section is used for the identification of the implementation and the client. It provides information about the implementation to be tested and the system it runs on, as well as the supplier of the implementation and/or the client and a contact person responsible for answering questions about the ICS or the product in general.

The fourth section of the ICS proforma identifies the relevant OSI specification to which the proforma applies, including the complete title and reference number. In addition, the client shall indicate which editions and/or versions of the specification are supported.

The fifth section contains another corrigenda table, which references the corrigenda to the OSI specifications which have been (or must be, this is not clear yet) taken into account in the implementation.

The last section of the ICS proforma is a questionnaire, hierarchically organized in tables, which is used to identify the capabilities supported by the implementation.

The ICS proforma will also ask whether all mandatory capabilities are implemented and will point out the consequences of a negative answer.

The following text is devoted to the capabilities and parameters listed in the ICS proforma tables for major capabilities, supported PDUs, supported parameters within the PDUs, timers, negotiation capabilities, protocol error handling and multi-specification dependencies. The structure of these tables reflects the structure of the static conformance requirements and the technical categories of capabilities.

ICS PROFORMA TABLES

The ICS proforma tables list all capabilities and options of the OSI specification. One row of a table shall cover one capability or option. The order of the columns of the table is given below, where "[...]" means optional and "?" means empty in the proforma and to be completed by the client.

Each proforma item consists of the following components:

- item number,
- item name,
- [OSI specification clause reference],
- one or more context groups,

where each context group consists of several columns:

- status,
- [condition predicate],
- [OSI specification clause reference],
- support?,
- [allowed values],

- [supported values?],
- [mnemonic],
- [user comments?].

Context groups are used to separate the information for different contexts like sending and receiving. The meaning of the column entries is as follows:

ITEM NUMBER assigns a reference number (usually 1,2,...) to each row or item of a table. It is used in other parts of the ICS proforma or in an abstract test suite for uniquely referencing individual answers of the ICS given in the support and supported values columns, when this information is required for specifying predicates, conditional expressions, test suite parameters (see 3.4.4) and test suite selection expressions (see 3.4.5). A complete reference to an ICS proforma answer is composed of the following components and given in the following sequence:

- identifier for the ICS proforma specification, which is required if the reference is made from outside the ICS proforma,
- number or smallest subclause of the relevant table, followed by /,
- item number or mnemonic reference to the item and
- /a, /b, ..., if more than one response is possible in a row to refer to single responses.

ITEM NAME will indicate the type of the item, e.g. protocol class or PDU, if the table lists the implemented protocol classes or PDUs. The item name itself will range over possible values for this type (e.g. Class0, Class1,... for the first example item and CR, CC, DT, ... for the second). Where appropriate, per row or per context group in a row, a reference to the relevant OSI specification clause is given.

STATUS gives the capability status of the item. If identical throughout the table it may be given once, outside of the table. Its possible values are m, M (for mandatory), o, O (for optional), c, C (for conditional), x, X (for prohibited, which used to be called excluded), n/a, N/A (for not applicable), i or I (for out-of-scope, which used to be called irrelevant) and o. (or O.) <integer> for the restricted optional cases, the same reference integer being used in a footnote explaining the restriction (e.g. "o.1: support of exactly one of these options is required").

PREDICATE specifies the predicates for conditional capabilities. It will be given in a form described below.

SUPPORT is the main column to be filled in, where Y, y, Yes or yes means that the capability is implemented and N, n, No or no means that this is not the case. Finally N/A, n/a or – means that no answer is required. Upper and lower case may be used freely. Alternatively to spaces in the support column for explicit answers by the client, separate tick boxes may be provided for each possible answer, and the client need only mark one of them for each item. An example with tick boxes is given in Table 2.1.

TABLE 2.1. An ICS proforma for major capabilities example

2.1 Major capabilities						
Item Number	Capability	Reference	Status	Support		Mnemonic
				Yes	No	
1	cap1	10.1	o.1	[]	[]	CAP1
2	cap2	10.2	o.1	[]	[]	CAP2
3	cap3	10.3	c1	[]	[]	CAP3
4	cap4	10.4	c1	[]	[]	CAP4

o.1:at least one of these capabilities has to be supported c1:IF CAP1 THEN o ELSE x

ALLOWED VALUES entry describes restrictions on type, length or range of values supported. These should cover the values or range to be provided by the client in the SUPPORTED VALUES entry.

MNEMONIC may be used to specify mnemonic identifiers as alternatives to item numbers for referencing purposes. Additional space should be provided for user comments or references to detailed comments outside of the table.

ICS PREDICATES

In order to avoid ambiguities, a notation for the predicates and conditions to be used for conditional capabilities is recommended. The ample presence of alternative notations again reflects the respect for existing divergent notations at the time of standardization.

Predicates in an ICS proforma are used to define under which circumstances a conditional capability is mandatory, optional, or prohibited. They are specified in one of the following forms:

- as an explicit ICS response reference, i.e. a full reference to a support column entry in the present ICS, in the format described in the previous section. For example, the predicate 2.1/1 evaluates to TRUE, if support is indicated in row 1 of Table 2.1, otherwise it evaluates to FALSE.
- as a predicate name, defined elsewhere in the ICS proforma, usually in a footnote to the table, with either
    - an explicit reference to an ICS proforma support answer, as for example "IUT_is_initiator = 2.3/1a" at the bottom of Table 2.2, or to

- a relational expression, as defined below, or to
- a predicate expression, as defined below, such as "¬IUT_is_initiator" in Table 2.6.

TABLE 2.2. Use of the predicate column

2.2 Protocol Phases					
Item number	Phase	Reference	Status	Predicate	Support
1	set-up	11.1	m		
2	data transfer	11.2	o		
3	disconnect	11.3	c: m	IUT_is_initiator	
			o	¬IUT_is_initiator	

IUT_is_initiator = 2.3/1a

A *relational expression* compares a full reference to a value column entry in the present ICS with the value of an expression, e.g. "2.4/1b < 64", where the entry reference evaluates to the integer written there. A precise grammar for relational expressions is not given.

A *predicate expression* involves logical connectives and predicate names. A precise grammar for predicate expressions is not given. It may be recommended to follow TTCN conventions in this case. Thus, predicate_3 could be defined by "predicate_3 = predicate_1 OR (predicate_2 AND (is_entry_lt_4))." The predicate name "is_entry_lt_4" could be defined by "is_entry_lt_4 = 2.4/1b < 4," i.e. involving a relational expression .

Conditional requirements are denoted by either of

- "c:" in the status column, followed by a list of status symbols, each on a separate line, and each followed in the predicate column by either a predicate name or an ICS response reference. The former may be preceded by a negation sign, written ^ or ¬. An example is shown for item 3 in Table 2.2. A permitted short form is used in Table 2.6.
- "c" in the status column, followed by an integer, under which label a status expression is defined elsewhere in the ICS proforma. Examples are given in Table 2.3.

Status expressions are IF-THEN-ELSE expressions, nested if necessary, which evaluate to an unconditional status. The status expression labels c1 in Table 2.1 and c2 in Table 2.3 and their expansion are examples for this definition method.

TABLE 2.3. Supported PDUs during connection set-up example

2.3 Supported PDUs in Phase Connection Set-up						
Item number	PDU	Reference	Sending Status	Support	Receipt Status	Support
1	CR	13.1	o		m	
2	CC	13.2	c2		c3	

c2: IF IUT_is_initiator THEN n/a ELSE m c3: IF ¬IUT_is_initiator THEN m ELSE n/a

TABLE 2.4. Supported parameters in PDU X example

2.4 Supported Parameters in PDU X						
Item number	Parameter	Reference	Status	Support	Values Allowed	Supported
1	Field1	13.4.1	o		0..128	
2	Field2	13.4.2	m		'01'H	

EXAMPLES OF PROFILE RL TABLES

Table 2.5 to Table 2.8 give some examples of profile RL tables that are related to the ICS proforma Tables 2.1 to 2.4. Table 2.5 indicates that the status of capability cap1 has changed to mandatory, while all other capabilities are irrelevant for this profile.Table 2.6 shows that data transfer and disconnect also have been made mandatory for the profile. Table 2.7 indicates that sending and reception of CR and CC PDUs have been made mandatory. Finally, Table 2.8 shows that the allowed value range of the parameter Field1 has been restricted to the single value 0.

TABLE 2.5. Profile RL table for major capabilities

2.5 Major capabilities					
Item Number	Capability	Reference	Protocol Status	Profile Status	Mnemonic
1	cap1	10.1	o.1	m	CAP1
2	cap2	10.2	o.1	i	CAP2
3	cap3	10.3	c1	i	CAP3
4	cap4	10.4	c1	i	CAP4

o.1:at least one of these capabilities has to be supported c1:IF CAP1 THEN o ELSE x

TABLE 2.6. A profile RL table for protocol phases

2.6 Protocol Phases				
Item number	Phase	Reference	Protocol Status	Profile Status
1	set-up	11.1	m	m
2	data transfer	11.2	o	m
3	disconnect	11.3	IUT_is_initiator: m ¬IUT_is_initiator: o	m

TABLE 2.7. A profile RL table for supported PDUs during connection set-up

2.7 Supported PDUs in Phase Connection Set-up						
Item	PDU	Reference	Sending		Receipt	
number			Protocol Status	Profile Status	Protocol Status	Profile Status
1	CR	13.1	o	m	m	m
2	CC	13.2	c2	m	c3	m

c2: IF IUT_is_initiator THEN n/a ELSE m c3: ¬IF IUT_is_initiator THEN m ELSE n/a

TABLE 2.8. A profile RL table for supported parameters in PDU X

2.8 Supported Parameters in PDU X						
Item number	Parameter	Reference	Protocol Status	Profile Status	Values Allowed By Protocol	Allowed By Profile
1	Field1	13.4.1	o	o	0..128	0
2	Field2	13.4.2	m	m	'01'H	'01'H

2.4.2 *The IXIT proforma*

As mentioned in Subsection 2.3.3, IXIT proformas are used to complement the information obtained from ICS proformas. They are related to abstract test suites, profile specific test specifications (PSTS) and the means of testing. Therefore, in [ISO 9646-5] only the skeleton structure of an IXIT proforma, independent of any OSI specifications, is provided in an informative annex. Without going into details of the different items within this IXIT proforma, we simply list them in the following tables, where the text in italics provides explanations and does not appear in a real IXIT proforma. Additionally the symbols • • • and – – –, which are also not included in real IXIT proformas, are used in Tables 2.9 to Table 2.12 to denote the information required from the test laboratory (indicated by • • •) and the information to be provided by the clients (indicated by – – –).

TABLE 2.9. Administrative part of the IXIT proforma

IXIT PROFORMA FOR	• • •	*Name of the abstract test suite or profile specific test specification*
1. IDENTIFICATION SUMMARY		
IXIT number:	• • •	
Test laboratory name:	• • •	
Date of issue:	• • •	
Issued to:	• • •	*Name of client*
The test laboratory may include client or contract references in the identification summary.		
2. ABSTRACT TEST SUITE SUMMARY		
Protocol specification(s):	• • •	*Protocol(s) to be tested*
Information object specification(s):	• • •	*Information object(s) to be tested,*
ATS specification:	• • •	*ATS or PSTS reference*
Abstract test method(s):	• • •	*ATM(s) used in the ATS or PSTS*
3. TEST LABORATORY		
Test laboratory identification:	• • •	*Name and addressing details*
Accreditation status of the test service:	• • •	
Accreditation reference:	• • •	
Test laboratory manager:	• • •	*Name*
Test laboratory contact:	• • •	*Name*
Means of testing:	• • •	*Identification of MOT, including version number*
Instructions for completion:	• • •	*Special instructions necessary for the completion and return of the proforma by the client.*
4. TEST CLIENT		
This section contains information for the identification of the client. Typically, this should include:		
Client identification:	– – –	*Name and addressing details*
Client test manager:	– – –	*Name*

TABLE 2.9, continued. Administrative part of the IXIT proforma

Client contact:	---	*Name*
Test facilities required:	---	*Particular facilities required for testing, if a range of facilities is provided by the test laboratory.*
5. SUT		
Name:	---	*System name*
Version:	---	*System version*
SCS reference:	---	
Machine configuration:	---	*Machine, on which the SUT will reside*
Operating system identification:	---	
Upper tester identification:	---	*if any*
Upper tester validation date:	---	*if appropriate*
IUT identification:	---	*Implementation name*
ICS reference for IUT:	---	
Limitations of the SUT:	---	*Information explaining if any of the abstract tests cannot be executed, for example, non-realization of ASPs if the Remote Test Method is used.*
Environmental conditions:	---	

6. ANCILLARY PROTOCOLS

This section is completed by the client in conjunction with the test laboratory. In the following table, the client identifies relevant information concerning each ancillary protocol of the SUT. One clause is required for each ancillary protocol included in the SUT other than the IUT itself. The information required is dependent on the MOT and the SUT, and covers all the addressing, parameter values, timer values and facilities as defined by the PICS for each protocol.

Protocol Name	Version No.	(opt.) PICS Ref.	(opt.) PIXIT Ref.	(opt.) PCTR Ref.
--- / •••	--- / •••	--- / •••	--- / •••	--- / •••

TABLE 2.10. Protocol information of the IXIT proforma

7. PROTOCOL INFORMATION FOR *Protocol Name*

This section is completed by the test laboratory and client in consultation. It applies only to PIXIT proformas.

7.1 PROTOCOL IDENTIFICATION

Specification reference:	--- / • • •	*Reference including edition*
Protocol version:	--- / • • •	*Protocol version number*
PICS reference:	--- / • • •	

7.2 IUT INFORMATION

This section should include such items as addresses, parameter values and timer values required to test the IUT. It should be a refinement of the information provided in the PICS for the relevant protocol, but not in conflict with it. Furthermore it should include test suite parameters where they are identified in the test specification.

7.2.1 ADDRESSES

This clause should identify the SAP addresses to be used:

--- *a) by the lower tester to access the IUT;*

• • • *b) by the SUT to access the lower tester.*

7.2.2 PARAMETER VALUES

The client provides the parameter ranges. The exact parameter value to be used is agreed between the client and the test laboratory.

Parameter name	Parameter type	PICS clause	Parameter range	Parameter value
• • •	• • •	• • •	---	--- / • • •
• • •	• • •	• • •	---	--- / • • •

7.2.3 TIMER VALUES

Timer name	Timer type	PICS clause	Timer range	Timer value
• • •	• • •	• • •	• • •	---
• • •	• • •	• • •	• • •	---

7.2.4 PROCEDURAL INFORMATION

This section deals with limitations of the IUT, which may result in abstract test cases that cannot be executed.

TABLE 2.11. Information object identification of the IXIT proforma

8 INFORMATION OBJECT IDENTIFICATION FOR	• • •	*Information object name*
This section applies only to information object IXIT proformas.		
8.1 INFORMATION OBJECT IDENTIFICATION		
Object identifier:	• • •	
Specification reference:	• • •	*Reference including edition*
ICS Reference:	• • •	
8.2 IUT INFORMATION		
This section should include such items as parameter values which are related to the information object and which are required to test the IUT. It should be a refinement of the information provided in the ICS for the relevant information objects, but not conflict with it. Furthermore, it should include test suite parameters where they are identified in the test specification.		
8.2.1 PARAMETER VALUES		
The client provides the parameter range and the exact parameter value to be used is agreed between the client and the test laboratory.		

Parameter name	Parameter type	PICS clause	Parameter range	Parameter value
• • •	• • •	• • •	– – –	– – – / • • •
• • •	• • •	• • •	– – –	– – – / • • •

8.2.2 PROCEDURAL INFORMATION

This section should identify requirements for testing, placed by the ATS or PSTS specification, which may not be realizable by the SUT and which may result in abstract test cases which cannot be executed.

TABLE 2.12. Profile Specific Information of the IXIT Proforma

9 PROFILE SPECIFIC INFORMATION FOR	• • •	*Information profile name*
This section applies only to profile specific IXIT proformas.		
9.1 PROFILE IDENTIFICATION		
Profile identifier:	• • •	Identifier from ISO/IEC TR 10000-2 taxonomy
Specification reference:	• • •	*Reference including edition*
Profile specific ICS reference:	• • •	
9.2 CONFIGURATION INFORMATION		
This section should include information on how the IUT is to be configured for testing its conformance to the identified profile. This should be specified in sufficient detail to allow the test campaign(s) to be reproduced with the same configuration of the SUT at a later date, if this should be necessary.		

2.4.3 The SCS proforma

The SCS gives an index of the relevant OSI specifications, ICSs, test reports and configuration informations, as well as the identification of the system and the client. The structure and the contents of an SCS proforma are shown in Table 2.13, whose format is not standardized.

Xrefs are cross-reference numbers or mnemonics. "Specification" is the OSI specification, e.g. the protocol specification. "Combinations" are the combinations of protocols and information objects that are supported by the SUT, regardless of whether these combinations belong to profiles or not. The configurability information can be given in other forms, which are also suggested in the standard.

TABLE 2.13. SCS proforma

ADMINISTRATIVE AND SYSTEM INFORMATION				
• *System (name of product, version number release date, etc.)*				
• *Supplier of the system*				
• *Client*				
• *Contact person*				
PROTOCOL INFORMATION				
Protocol name	Specification reference	PICS reference	PCTR reference	Xref
•••	•••	•••	•••	•••
INFORMATION OBJECT INFORMATION				
Information object name	Specification reference	Information object ICS reference	PCTR reference	Xref
•••	•••	•••	•••	•••
PROFILE INFORMATION				
Profile identifier	Specification reference	Profile specific ICS reference	SCTR reference	Xref
•••	•••	•••	•••	•••
COMBINATION INFORMATION				
Combination identifier	Profile Xref	List of protocol Xrefs		List of information object Xrefs
•••	•••	•••		•••
CONFIGURABILITY INFORMATION				
Combination identifier	Always available?	Available by dynamic reconfiguration?		Available by static reconfiguration?

TABLE 2.13, continued. SCS proforma

•••	•••	•••	•••
CONFORMANCE ASSESSMENT INFORMATION			
Base or profile testing?	Profiles to be tested (Xrefs)	Protocols to be tested (Xrefs)	Information objects to be tested (Xrefs)
•••	•••	•••	•••

2.4.4 Profile testing documents

In this section the remaining profile documents are presented. These are the profile test specification summary, the profile specification document and the profile test specification. Before we present schemas for these documents, we give definitions of technical terms of protocol profile testing:

- A *component of a profile* is a collection of one or more protocols and, possibly, information objects which have to be jointly tested.
- A *common subprofile* is a part of a profile which can be used by a number of other profiles. It may refer to other common subprofiles. If a common subprofile is a complete profile, it is called a *common profile.*
- The *PSTS test case replacement list* identifies the abstract test cases of the related ATSs that have to be deselected or replaced by test cases defined in the profile specific test specification.
- The *profile specific test specification* (PSTS) contains the modified or additional profile-specific test cases.

PTS-SUMMARY

The *profile test specification summary* (PTS-summary) provides references to all documents required for profile testing. It is used as an introduction to the profile test specification. The structure and contents of the PTS-summary are shown in Table 2.14.

PROFILE SPECIFICATION DOCUMENT

The *profile specification document* (PSD) contains one or more profile specifications and their requirements lists, their PTS-summaries and the PSTS.

As mentioned, the development of a profile test specification is based on existing test specifications of the related base standards and makes use of the related TSS&TP, ICS, IXIT and ATS documents.

TABLE 2.14. PTS summary

Section 1: Profile identification
profile identifier according to the TR-10000 classification scheme reference to the profile specification reference to the profile RL reference to the PSTS reference to the profile specific ICS proforma reference to the partial profile XRL reference to the partial profile specific IXIT proforma reference to an SCS proforma for the profile comments
Section 2: Components of the PTS
summary of Subsections 2.1 to 2.n
Subsection 2.1 - 2.n: Conformance testing specifications for the 1st to the nth component
complete protocol identification(s) complete information object identification, if applicable reference to information object ICS proforma(s) reference to the PICS proforma reference to the TSS&TP specification reference to the PSTS, if additional test purposes are needed
Subsection 2.1.1 - 2.n.m: Use of 1st to m^{th} ATS for 1st to n^{th} component
reference to ATS specification additional protocols, services, and information objects information name of abstract test method reference to test management protocol, if applicable reference to partial PIXIT proforma reference to PSTS , if additional test cases are needed
PTS-summary for profiles which reference common profiles
reference to PTS-summary
Section 3: Test specifications related to more than a single profile component
reference to PSTS and relevant section about additional test suite structure reference to PSTS and relevant section about additional test purposes reference to PSTS and relevant section about additional test cases
Section 4: Conformance clause
text for conformance clause from annex A of [ISO 9646-6],

The layout shown in Table 2.14 is not standardized. In principle, a PTS does not represent a full test specification, comparable to an ATS, because it mainly provides references to other test documents. Only the PTS summary and the PSTS are published as separate parts of the profile specification document. Note that a profile specification may contain more than one profile. A complete profile specification for a single profile typically comprises

- in part 1
  - the profile specification,
  - the profile RL, and
  - the profile specific ICS proforma,
- in part 2 the PTS summary, and
- in part 3 the PSTS.

A complete profile specification for more than one profile typically comprises three subspecifications,

- the first containing for each profile a separate part, providing
  - the profile specification,
  - the profile RL, and
  - the profile specific ICS proforma,
- the second containing the respective PTS summaries on corresponding parts, and
- the third containing the respective PSTSs in corresponding parts, plus, if applicable, a PSTS common part.

PROFILE TEST SPECIFICATION

The *profile test specification* (PTS) is a set of documents relevant for profile testing. The PTS includes the PTS-summary, the TSS&TPs and the ATSs of the related protocols, and optionally the PSTS. The production of a profile test specification includes the following steps:

- analysis of the profile ICS and profile IXIT, including ICS and IXIT proformas of the related base specifications and the profile RL, as well as reporting of possible defects to the base specification defining groups,
- analysis of the TSS&TP and ATS of the related base specifications and reporting of possible defects to the base specification defining groups,
- analysis of the ATS parameterization and selection specification tables, and possibly their completion and modification,

- specification of new profile specific test purposes and test cases, and standardization of the PSTS,
- replacement of test purposes and test cases according to the PSTS test case replacement list, which identifies the relevant test cases,
- collection of references to all relevant test specifications in the PTS-summary.

The profile testing methodology can be applied to different kinds of profiles. In special cases, internationally agreed and harmonized documents which define a profile or a group of related profiles are denoted as international standardized profiles (ISPs). Selected parameters and options of ISP implementations, and additional constraints placed on possible answers in base standard PICSs, are covered by documents called

- International standardized profile implementation conformance statement (ISPICS) and
- International standardized profile requirements list (IPRL).

Similarly, the profile IXIT of an ISP is called ISPIXIT. This is analogous to the more general terms profile ICS, profile RL and profile IXIT.

2.5 Abstract test methods

OSI implementations can be integrated into their system environment in many different ways. The possible configurations differ in their structure and in the observability of the IUT's behavior during testing. The OSI testing methodology, therefore, defines a set of *abstract test methods* (ATM), each describing an abstract test configuration, possible types of observation and control, and providing for test coordination. The ATMs are considered to be sufficient for the entire spectrum of commercially available implementations and systems under test. ATMs are determined

- by the configuration of the IUT with respect to the SUT,
- by the total testing configuration, encompassing the test system, the system under test, and the underlying service provider,
- by the points in the total testing configuration at which the tester can control and observe the IUT, and
- by the different degrees of precision of the test coordination procedures between the tester parts.

In this section, we will outline the CTMF view of test methods.

2.5.1 A classification of systems and implementations

From the viewpoint of OSI testing, CTMF classifies systems according to three criteria:

- whether the system is an end system or a relay system,
- in which layers it uses OSI protocols, and
- in which layers it alternatively uses non-OSI protocols.

Although it seems that the third criterion is not really used in CTMF, we render the list of system configuration patterns that are identified in the standard, cf. Figure 2.3:

- *Complete open end systems* use OSI protocols in all OSI layers.
- *Partial open end systems* use OSI protocols in all layers from the physical layer up to another layer below the application layer to provide at least an OSI service. The layers above the partial open end system may be non-OSI layers or even empty.
- *Open relay systems* use either OSI protocols in the physical, data link and network layers (open network relay system) or in all OSI layers (open application relay system).
- *Combined systems* can be constructed from the above three basic configurations which use coexistent OSI and non-OSI protocols above an OSI service. But for the purpose of OSI testing, this pattern can be considered equivalent to the second configuration, i.e. partial open end system.

The objects of OSI conformance testing are *implementations under* test (IUT), implementations of one or a combination of OSI protocols from one layer or several adjacent layers. These protocols should be related, i.e. each one of them uses the service provided by another or provides the service used by another, or both. In a complete or a partial open system, the IUT is accordingly called a *single-protocol IUT* or *multi-protocol IUT*. IUTs in open relay systems are not categorized that way but are said to include the relay function. Hence, the classification of systems not only depends on the criteria given above, but also on the layers spanned by the IUT.

The demarcation of the IUT within the SUT is subject to an agreement between the client and the test laboratory. It is described in the ICS and IXIT documents.

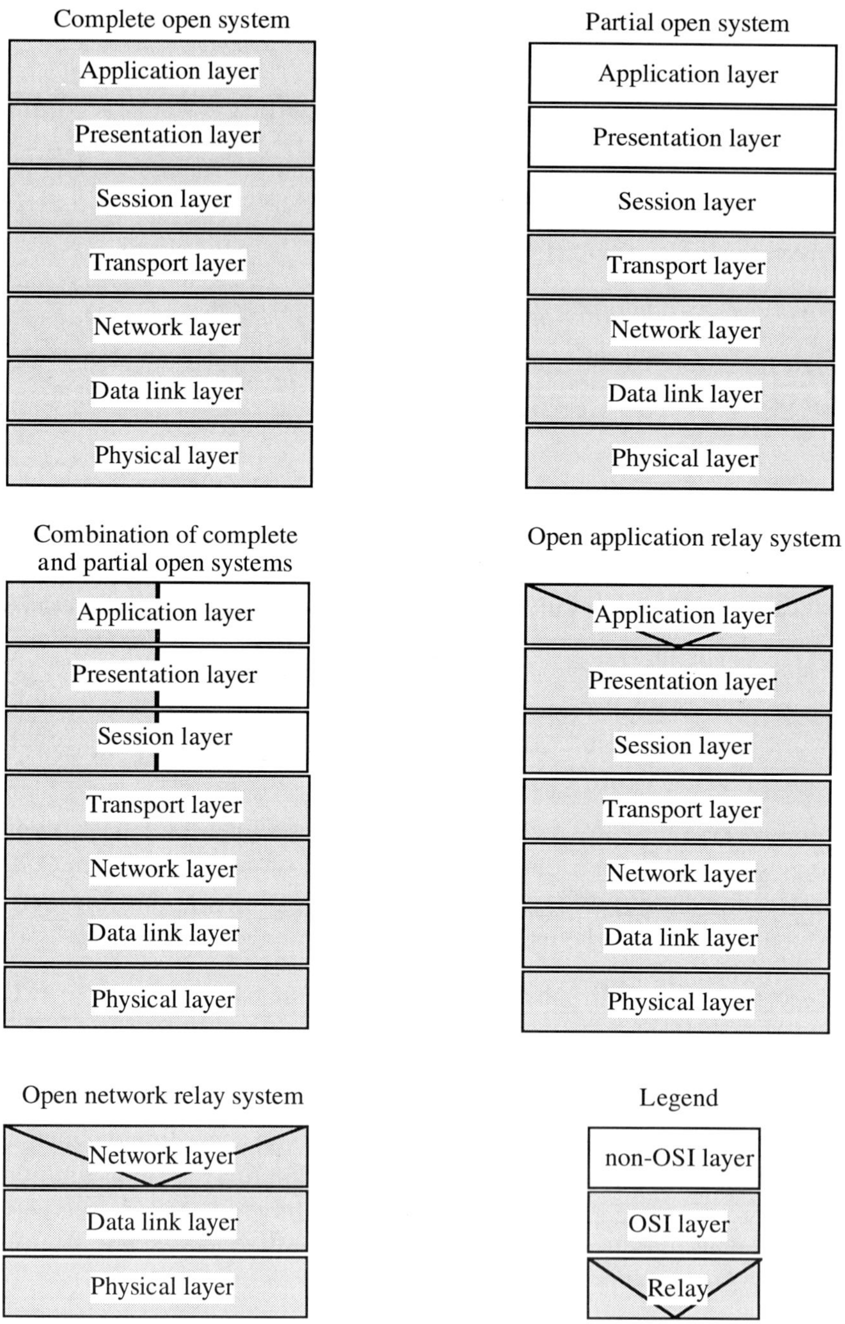

FIGURE 2.3. Classification of systems

2.5.2 Basic testing concepts

OSI testing concepts are based on the OSI reference model. Abstract test methods are characterized by the points, nearest to the IUT, where the tester can observe the IUT, i.e. receive input from it, and control the IUT, i.e. submit input to it. The combination of observation and control events constitutes the behavior of the IUT, as perceived by the tester. The behavior of a protocol entity (e.g. the IUT) is defined in terms of protocol data units (PDU) and *abstract service primitives* (ASP) at its upper or lower boundary. The permissible range of behavior is defined by the dynamic conformance requirements of the protocol specification. CTMF uses the term abstract service primitives rather than service primitives to emphasize the symbolic nature of these primitives, which need not be explicitly implemented. Figure 2.4 illustrates these events as the only things observable about a black-box IUT. CTMF uses the term *conceptual testing architecture*. It is not clear in CTMF whether the configuration shown is meant to be the (one and only) conceptual testing architecture or just a conceptual testing architecture (of many possible ones) in an informal sense. At any rate, it is not considered as a standardized abstract test method, because it requires tester code in the SUT. These types of requirements on SUTs are avoided by CTMF, as real products cannot be forced to add code for conformance test purposes. Responders are the one exception, and even they are used carefully and in a limited fashion.

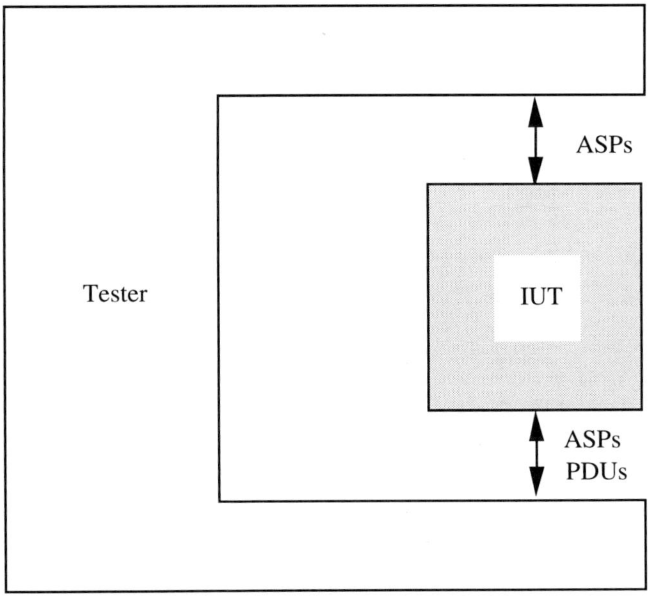

FIGURE 2.4. Conceptual testing architecture

2.5.3 *Points of control and observation*

It is assumed that test events, i.e. ASPs or PDUs, are observed (received from the IUT) or controlled (sent to it) and that these test events happen at service boundaries within the SUT or in the test system. The possible locations for these *points of control and observation* (PCO) are symbolically shown in Figure 2.5.

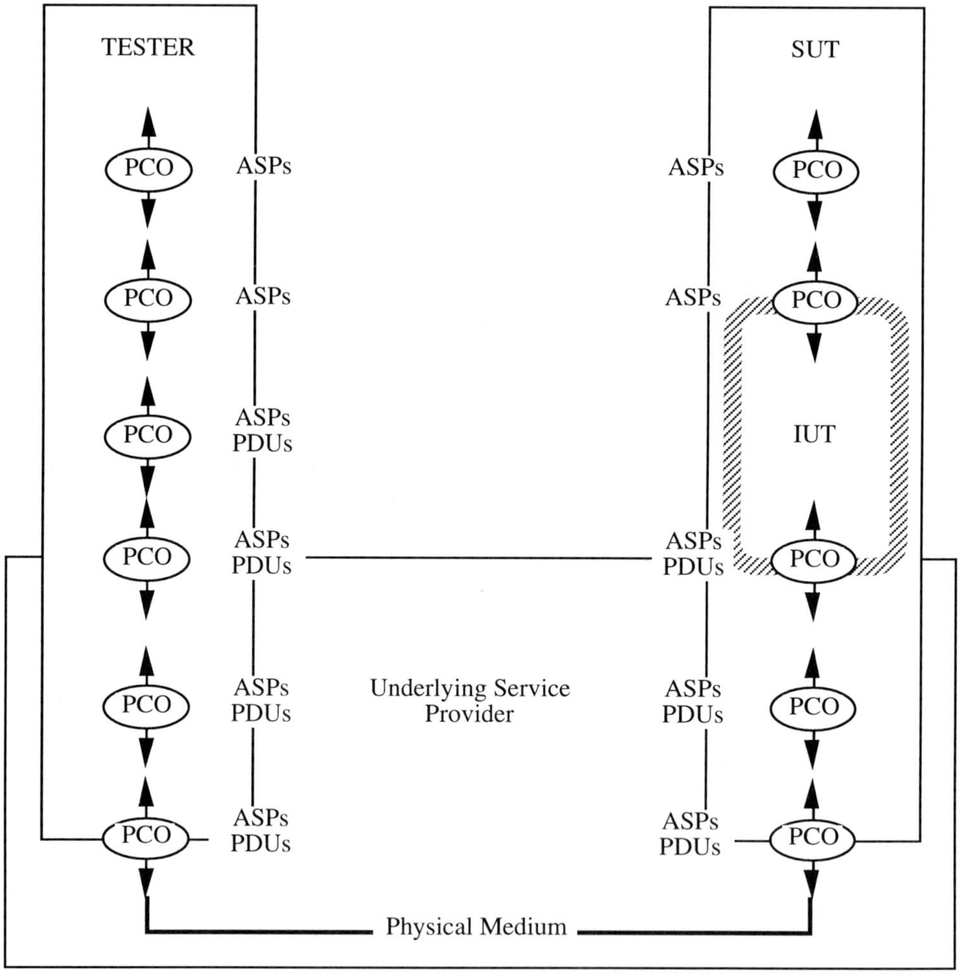

FIGURE 2.5. Possible positions of PCOs

They can be classified as at or above the upper IUT boundary, at or below the lower IUT boundary, or in between, and also in the SUT or in the test system. The abstract test methods preselect some useful combinations of PCO locations, which, of course, determine the ability to control and observe the IUT. The choice of a PCO does not fully impose a choice between ASPs and PDUs as the primary test events to observe and control. This choice seems to be left to the test suite writer. However, parts 1 and 3 of 9646 slightly disagree in this respect. Of course, the handling of ASPs includes the handling of the PDUs contained therein.

According to 9646, PCOs can be viewed as two queues, one for the test events to be sent to the IUT, and one for the test events received from the IUT. Note that there are some problems related to this picture. We will take a closer look at these in connection with the presentation of TTCN behavior descriptions in Chapter 4.

Commercially available IUTs ordinarily do not provide access below the lower IUT boundary in the SUT. Therefore, test events at or below the IUT's lower boundary can only be perceived, via the common underlying provider, by a remote peer entity in a test system. The service provider is assumed to be functioning correctly and sufficiently reliably to permit this. The necessary confidence in the service provider can be gained by prior testing of the lower layers. On the other hand, the realization of service boundaries is not a standardized requirement on protocol implementations. For that reason, the IUT's upper boundary also may be completely inaccessible for testing purposes, i.e. normally there will not even be any remote access to it. The presence or absence of access above the IUT is one of the features by which abstract test methods differ from one another.

2.5.4 Testing contexts

Since the completion of version 1 of 9646, the OSI conformance testing methodology has been expanded towards a multi-party testing methodology, providing two different testing schemes which are called the *multi-party testing context* (MPyT) and the *single-party testing context* (SPyT). MPyT and SPyT are defined as follows:

- The single-party testing context is relevant for testing an IUT which communicates with exactly one real open system, represented by a single lower tester.
- The multi-party testing context is relevant for testing an IUT which concurrently communicates with multiple real open systems, represented by more than one lower tester.

In the multi-party testing context, an IUT can be logically divided into several parts, called *IUT components*, where each IUT component communicates with a different lower tester. The IUT components may be coordinated by a special IUT component, as depicted in Figure 2.6.

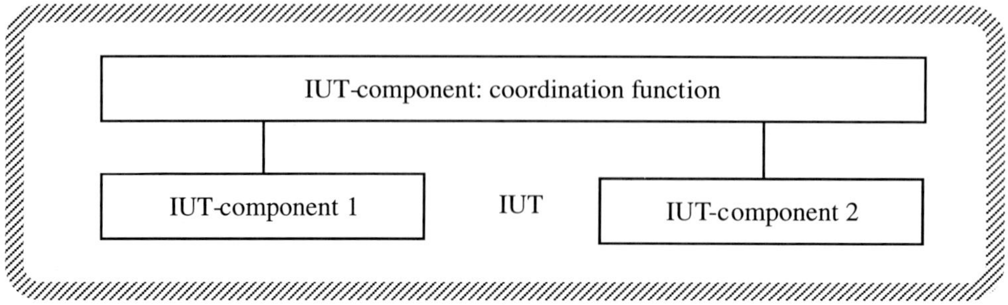

FIGURE 2.6. IUT components

Testing configurations in which all IUT components (except the coordination function) handle the same protocol are called *homogeneous testing configurations*. If, however, different protocols have been realized in at least two IUT components, then the configuration is called a *heterogeneous testing configuration*. In addition, it is allowed that different IUT components have different service access points to the same, or even to different, underlying service providers.

2.5.5 Testing functions

In the definitions of abstract test methods the following four principal tester roles, called abstract testing functions, are distinguished (see also Figure 2.7, which shows, by way of example, two lower and two upper testers. Note, that here, and in all following figures PCOs are indicated by solid circles):

- A *lower tester* (LT) controls and observes the IUT's lower service boundary. It does so indirectly, via the underlying service provider. In the SPyT context, the LT behaves as the peer entity to the IUT, and is responsible for the assignment of verdicts for the test cases. In the MPyT context, several lower testers working in parallel, called *parallel lower testers* (PLT), act as peer entities to the protocol entity or entities of the IUT.
- The *lower tester control function* (LTCF) has the task of coordinating all lower testers and assigning the test case verdicts. The use of an LTCF is mandatory in the multi-party context. In the single-party testing context, coordination is inapplicable.

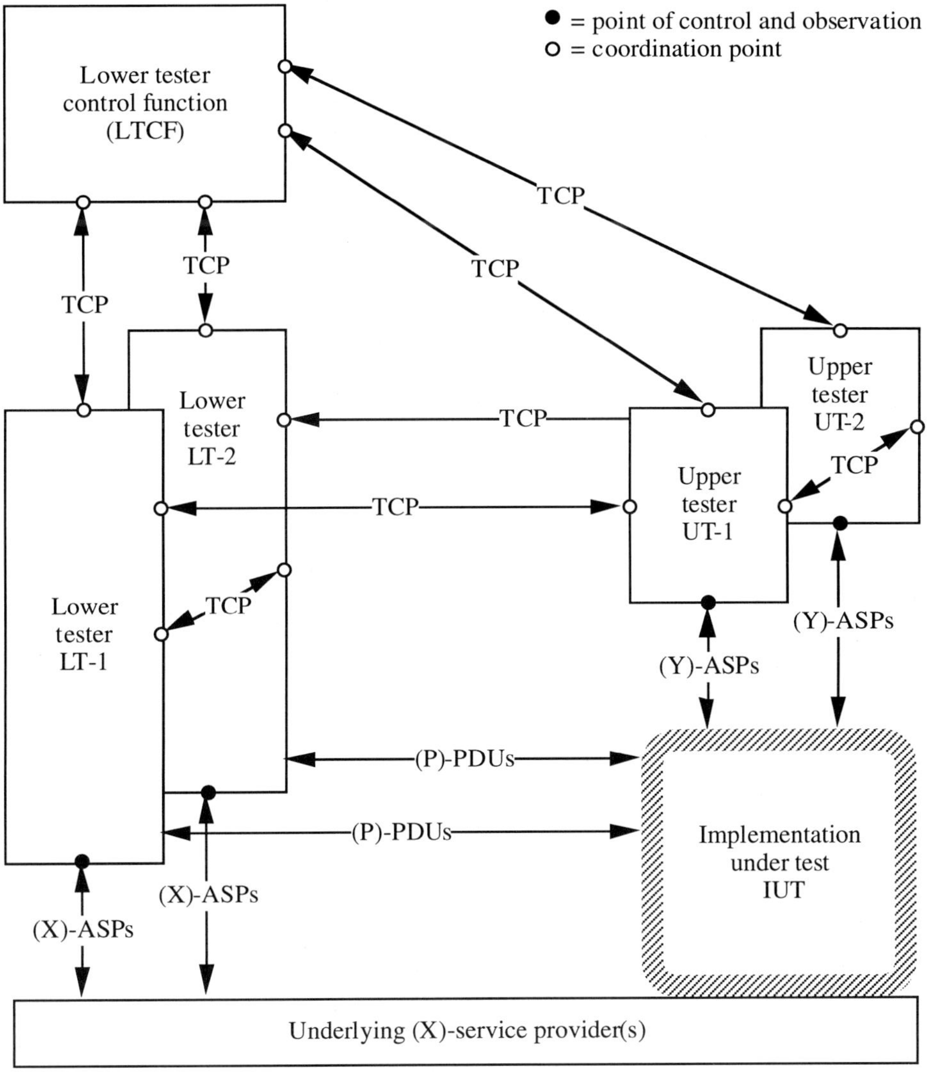

FIGURE 2.7. Abstract testing functions

– An *upper tester* (UT) controls and observes the IUT's upper service boundary, as defined by the chosen abstract test method. It does so either by operator access, via an application program interface, or via a hardware interface. For lack of standardized application program or hardware interfaces, operator intervention remains the only way for controlling and observing the upper service boundary of the IUT. In the SPyT context, the UT behaves as a user of the IUT. In the MPyT context, optional upper testers working in parallel, called *parallel upper testers*, act as users of the IUT.

- In the SPyT context, the *test coordination procedures* (TCP, TCPs) are required to ensure cooperation between the upper and the lower tester. In some test methods, the TCPs are defined by means of an explicit *test management protocol* (TMP), while in others requirements on the TCPs can only be inferred from the description of the TTCN operational semantics. In these cases, nothing is said about the realization of the TCPs themselves. Usually the LT will control the activities of the UT, being continuously informed by the latter about all send and receive events at the UT PCO. In the MPyT context, the test coordination procedures are also used to ensure cooperation between LTs and UTs.

2.5.6 *The classification of abstract test methods*

The example testing architecture shown in Figure 2.7 is to be understood as a general model for deriving possible abstract test methods. The different methods depend on the number and position of test components, abstract testing functions and PCOs. As mentioned before, IUTs themselves are classified according to the protocol or stack of protocols they include, as well as the services they provide at their upper, and use at their lower, boundaries.

Since the boundaries of an IUT need not match OSI layer boundaries, and a layer may comprise more than one protocol, CTMF uses a notation with "(X)-ASP," "(Y)-ASP" and "P_1 to P_n" as a generalization of the (N-1) and (N) notation of the OSI basic reference model [ISO 7498]. In the case of a single-protocol IUT spanning a complete OSI layer, (X)-ASPs correspond to (N-1)-ASPs and (Y)-ASPs correspond to (N)-ASPs. The lowest to highest protocol layer of a multi-protocol IUT are numbered symbolically P_1 to P_n. The underlying service below the protocol P_1 is called the (X)-service and similarly the service provided by the protocol P_n is called the (Y)-service. The same numbers apply for the respective layers in the LT.

To give an example in which the (N)-notation is insufficient, let us consider a CCR (Commitment, concurrency and recovery) implementation, whose "lower boundary events," the X-service ASPs, comprise both presentation and ACSE (Association control service element) ASPs. The ACSE service is not an OSI layer boundary, nor is the Y-service, i.e. the CCR service.

In single-protocol IUTs, P_1 coincides with P_n and is simply denoted by P. In all abstract test methods, the lower testers communicate with the IUT via an underlying service provider. The underlying provider of the physical layer is the physical medium. Note, that 9646 does not fully apply to the physical layer. For example, it cannot be used to test electrical characteristics. In abstract test suites, test events at the lower tester PCO will be specified in terms of (X)-ASPs and/or (P_1- to P_n)-PDUs. Alternatively, it is possible to specify events only by PDUs, though this may imply a reduced controllability or observability of the IUT. The specification of a broad spectrum of test configurations is possible for the single-party and multi-party testing context.

ABSTRACT TEST METHODS FOR MULTI-PARTY TESTING

The most general model in the multi-party testing context describes test configurations which use

- several parallel lower and upper testers (see Figure 2.7).

In special cases, there is

- only one upper tester (omit UT-2 in Figure 2.7), or even
- no upper tester at all (omit both UT-1 and UT-2 in Figure 2.7). The special case of this configuration with two LT's is appropriate for testing relay systems.

ABSTRACT TEST METHODS FOR SINGLE-PARTY TESTING

The most complex model in the single-party testing context, called the *multi-user variant,* describes test configurations with one lower tester and more than one upper tester (omit LT-2 in Figure 2.7). The second and most common model represents configurations with a single lower tester and a single upper tester (see Figure 2.7, with only LT-1 and UT-1). Finally, the third model (see Figure 2.7, with only LT-1) illustrates the simplest type of a single-party testing configuration, which does not use any upper tester at all.

Four abstract test methods are provided for end-system IUTs for the SPyT context, denoted as *local, distributed, coordinated* and *remote* test methods. They are mainly distinguished by the different assumptions about the existence and the location of the single upper tester. For unknown reasons, the multi-user variant is not an abstract test method.

A test method has the attribute *non-embedded,* if access to the upper service boundary of an IUT, on top of the protocol to be tested, is possible. On the other hand, it is called *embedded,* if control and observation is only possible at the uppermost service boundary of additional protocols on top of the IUT.

The non-embedded variant can be applied to single-protocol IUTs as well as to multi-protocol IUTs, the latter being tested as a whole, including (P_1)- to (P_n)-PDUs. Embedded test methods are used for a protocol layer P_e ($1 \leq e < n$) under test in a multi-protocol IUT which includes higher layers P_{e+1}, P_{e+2}, ..., P_n The complete set of test events is formed by (P_e)-, (P_{e+1})-, ..., (P_n)-PDUs and (P_{e-1})-ASPs. Usually, multi-protocol IUTs are tested *incrementally*, i.e. sequentially applying the embedded variant. In this approach, protocol P_1 is tested first, embedded under protocols P_2 to P_n. Next, P_2 is tested embedded under protocols P_2 to P_n, etc. Finally, the highest protocol, P_n, is tested in the non-embedded variant of the test method. The notion of testing more than one layer of multi-protocol IUTs at once was abandoned in CTMF for end systems after the difficulties in writing test suites including PDUs of more than one layer became obvious.

THE LOCAL TEST METHOD

The *local test method*, cf. Figure 2.8, is characterized as follows:

- It uses one PCO below the LT and another PCO below the UT at the upper service boundary of the IUT. The UT PCO is supposed to be a standardized hardware interface. The interface specification has to describe in particular the realization of the relevant ASPs.
- The test events at the LT PCO comprise (X)-ASPs and/ or (P_1- to P_n)-PDUs.
- The test events at the UT PCO comprise (Y)-ASPs.
- The UT is located in the test system. Hence the TCP requirements are realized locally within the test system.

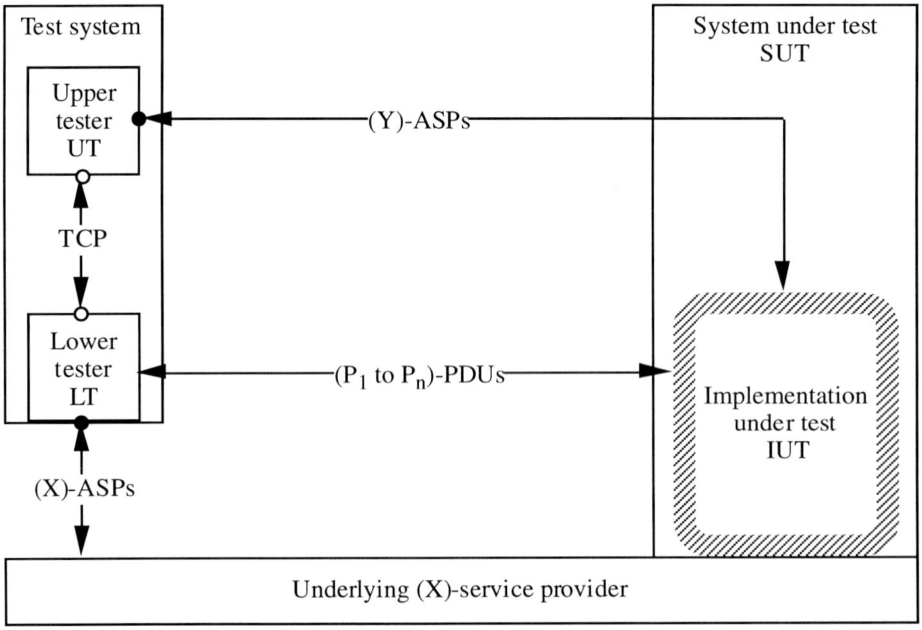

FIGURE 2.8. The local test method

It should be mentioned, that up to now abstract test suites based on the local test method do not exist.

THE DISTRIBUTED TEST METHOD

The *distributed test method*, cf. Figure 2.9, is characterized as follows:

- It uses one PCO below the LT and another PCO below the UT at the upper service boundary of the IUT. The UT PCO is supposed to be either a human user interface or a standardized programming language interface. In the latter case, the UT is located within the SUT and test suites cannot require anything but what the standardized interface specification provides.
- The test events at the LT PCO comprise (X)-ASPs and/ or (P_1 - to P_n)-PDUs.
- The test events at the UT PCO comprise (Y)-ASPs.

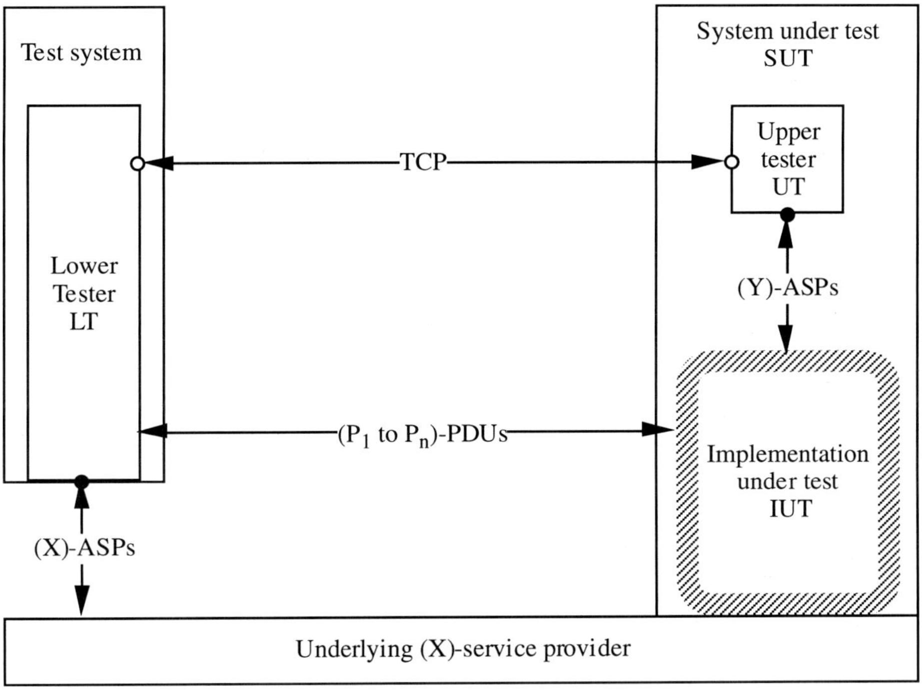

FIGURE 2.9. The distributed test method

- The TCP requirements specified in the ATS shall be met either by the human SUT operator or, in the case of the programming language interface, by the LT and UT together. Until standardized application programming interfaces exist, the distributed test method is restricted to applications with operator interfaces.

Since the local and the distributed test methods use the same PCOs, ASPs and PDUs, it should be possible to use identical or at least very similar abstract test suites for both. Only the TCP realizations and the UT access must differ between the two.

THE COORDINATED TEST METHOD

The *coordinated test method*, cf. Figure 2.10, is characterized as follows:

- It uses one PCO, below the LT. Tester access to the upper service boundary of the IUT is not required (although the standard seems to falter in this respect).
- The ability to observe and control the upper IUT interface indirectly is given by the UT located in the SUT, on top of the IUT, implementing the TMP (see below). It may achieve the effects on the IUT which could be achieved if service primitives at an upper service boundary of the IUT were available. Likewise, it may report via the TMP what could be observed in that case.
- The test events at the LT PCO comprise (X)-ASPs and/ or (P_1- to P_n)-PDUs. (Y)-ASPs are not explicitly used in the ATS.
- The TCP requirements are specified in the ATS by referring to a standardized *test management protocol* (TMP). The TMP is usually designed specifically for the protocol to be tested. Different protocols may have different TMPs. The TMP may, but need not, refer to (Y)-ASPs.

A test suite for the coordinated method may be supplemented by particular test cases for the validation of the correct functioning of the UT with respect to the TMP. TM-PDUs are exchanged between the LT and UT either embedded in the user data field of PDUs or via a second connection.

Each employment of a TMP in a conformance assessment process requires the filling of a TMP implementation statement, similar to the PICS, and including at least an entry for each TM-PDU.

THE REMOTE TEST METHOD

The remote test method, cf. Figure 2.11, is characterized as follows:

- It uses one PCO, below the LT. Tester access to the upper service boundary of the IUT is not required.

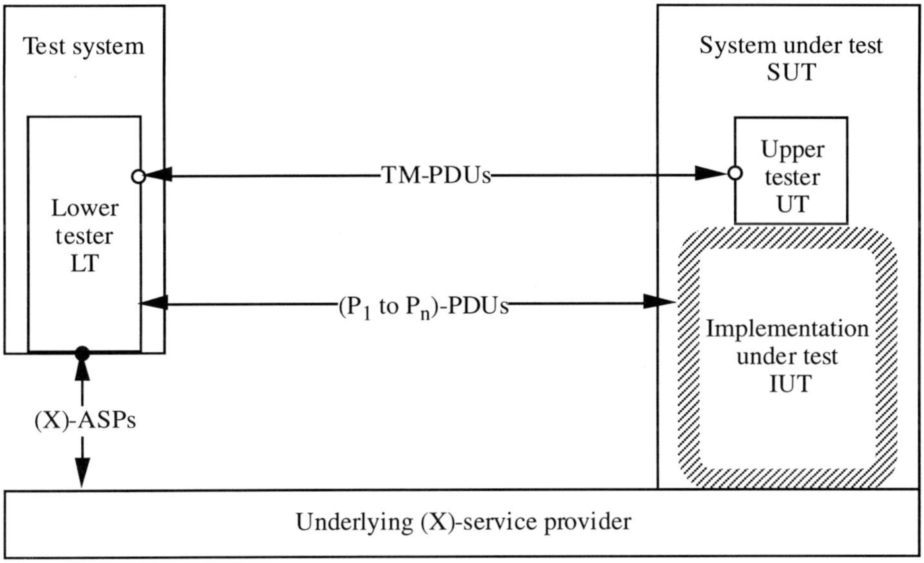

FIGURE 2.10. The coordinated test method

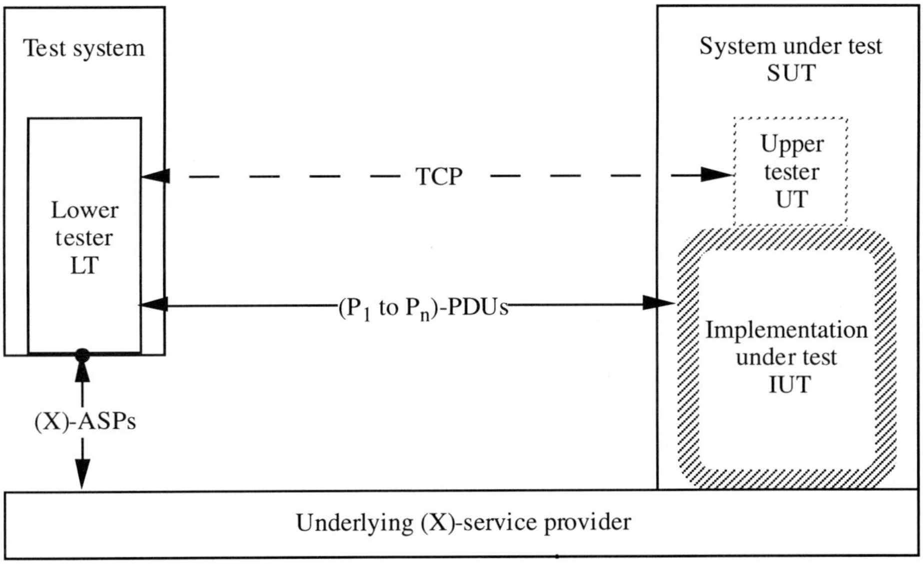

FIGURE 2.11. The remote test method

- The test events at the LT PCO comprise (X)-ASPs and/ or (P_1- to P_n)-PDUs. (Y)-ASPs are not explicitly used in the ATS.

- Some requirements for upper tester functions and test coordination procedures may be expressed in the ATS informally or implicitly. By a technique of "implicit specification" the ATS may ask the SUT (or rather its operator) at certain points to do whatever is necessary to make the IUT send PDUs that an UT could cause the IUT to send. However, the client may refuse, or be unable, to realize these informal or implicit requirements. This will be noted in the PIXIT, and the respective test cases will then have to be deselected. The consequences of a client's deselecting an excessive number of test cases on this account are not addressed in CTMF.

Note that even if all requirements for implicit specification, and hence for controllability of the IUT, are fulfilled, the lack of an upper tester prevents the tester from observing the IUT's upper interface.

2.5.7 Examples of testing configurations

Some examples for the use of the multi-user variant in single party testing are protocols which involve multiplexing (transport, class 2) or broadcasting of delivered messages (MHS). Examples for the use of the multi-party testing context are relay systems or transaction processing (TP) systems.

MHS TESTING CONFIGURATION

Specific aspects of MHS require IUTs to be tested in the multi-user variant of SPyT, for example if more then one recipient is connected to the IUT. The testing of such an MHS system (see Figure 2.12) includes the verification of the IUT's ability to copy messages when required, to transmit them to the relevant recipients, and to collect their expected responses. In this testing configuration as many UTs as recipients are required.

MULTIPLEXING TESTING CONFIGURATION

Multiplexing aspects of transport (or network layer) protocols also require IUTs to be tested in the multi-user variant of SPyT, for example if more than one transport (or network) connection is multiplexed over a single network (or data link) connection. In this testing configuration (see Figure 2.13) as many UTs as transport (or network) connections are required.

RELAY TESTING CONFIGURATION

The testing of relay systems requires IUTs to be tested in the multi-party testing context without using UTs. In Figure 2.14, a relay testing configuration is shown, where the relay system is connected to two different underlying service providers.

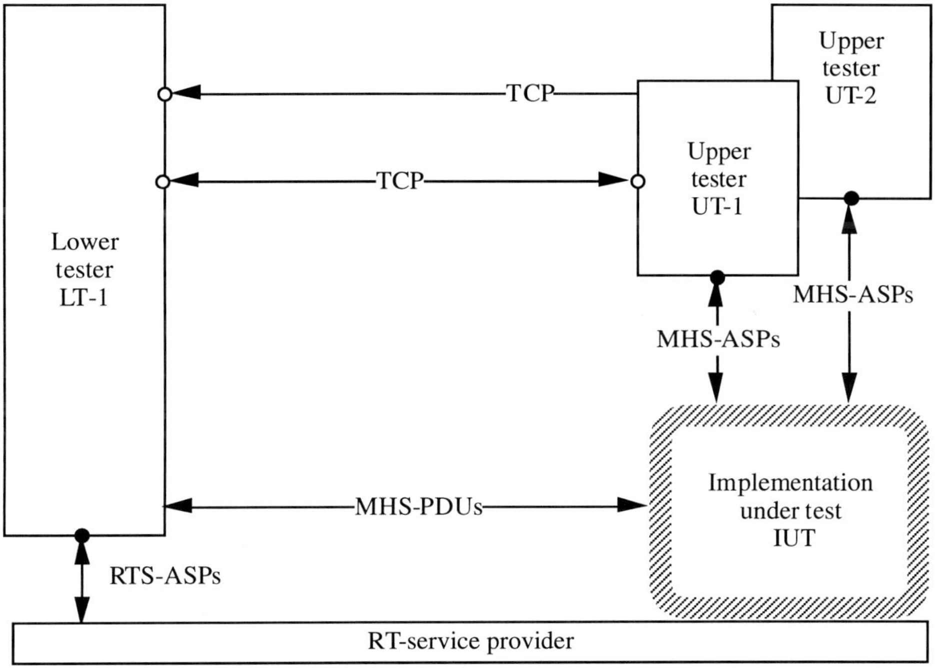

FIGURE 2.12. Multi-user testing configuration for MHS

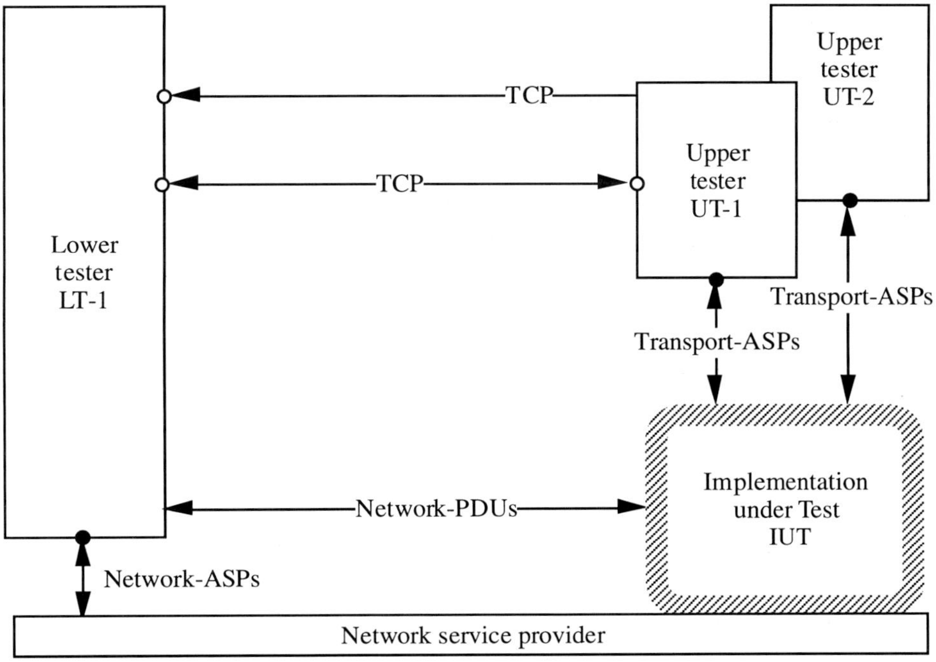

FIGURE 2.13. Multi-user testing configuration for multiplexing

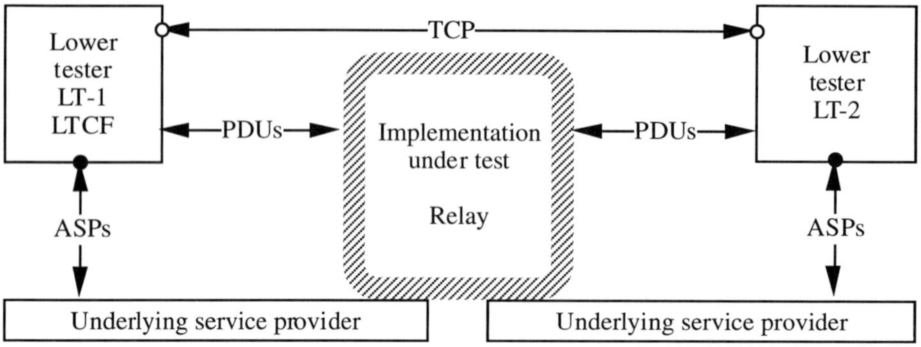

FIGURE 2.14. Multi-party testing configuration for relay systems

TP TESTING CONFIGURATION

Transaction trees are used in modelling *transaction processing* (TP). Three types of TP nodes are distinguished, which fulfil different tasks and which have to be tested accordingly. A possible transaction tree is shown in Figure 2.15.

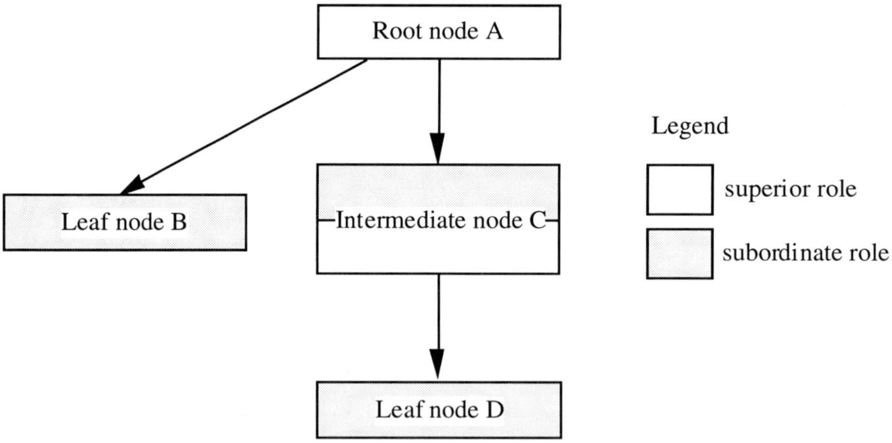

FIGURE 2.15. Transaction tree

The different types of TP nodes play the following roles:

- The *root node* represents the initiator of a transaction. It is superior to all nodes of a transaction and it is responsible for initiating transactions or operations which may influence individual sub-trees or even the whole transaction tree.

- An *intermediate node* operates in two roles, namely as a *superior node* and as a *subordinate node*. In the role of a subordinate node, an intermediate node receives transaction requests from a superior node, and in the role of a superior node, it delegates requests to its subordinate node, if it is not able to complete the transaction task by itself.
- A *leaf node* only operates in the role of a subordinate node.

Multi-party testing methods are required for transaction processing, especially if complex transaction trees have to be checked. However, it should be mentioned that transactions can be reduced to two-party communication between neighboring nodes. Therefore, whenever the coupling mechanisms are not of relevance, individual transaction branches can also be tested in the single-party testing context. In the first test configuration (Figure 2.16), the IUT is tested as a root node, i.e. exclusively in the role of a superior node. Depending on the specific application, two or more lower testers (LT-1, LT-2, ...) are required, which operate as subordinate TP-nodes.

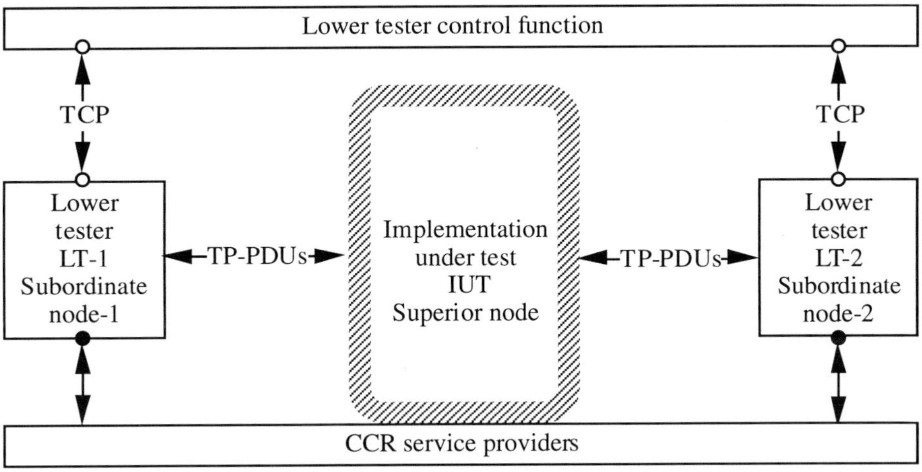

FIGURE 2.16. TP test configuration 1 – IUT in the role of a superior node

In the second test configuration (Figure 2.17), the IUT is tested as an intermediate node, i.e. in the roles of a superior and of a subordinate node. One lower tester (LT-1) operates as the root node, passing requests to the IUT, which can be either accepted or rejected locally. If a transaction request is accepted by the IUT, the IUT will perform the required actions in the role of a subordinate node, and will then act as a superior node and pass requests to its subordinate nodes, represented by the lower testers LT-2, LT-3, etc.

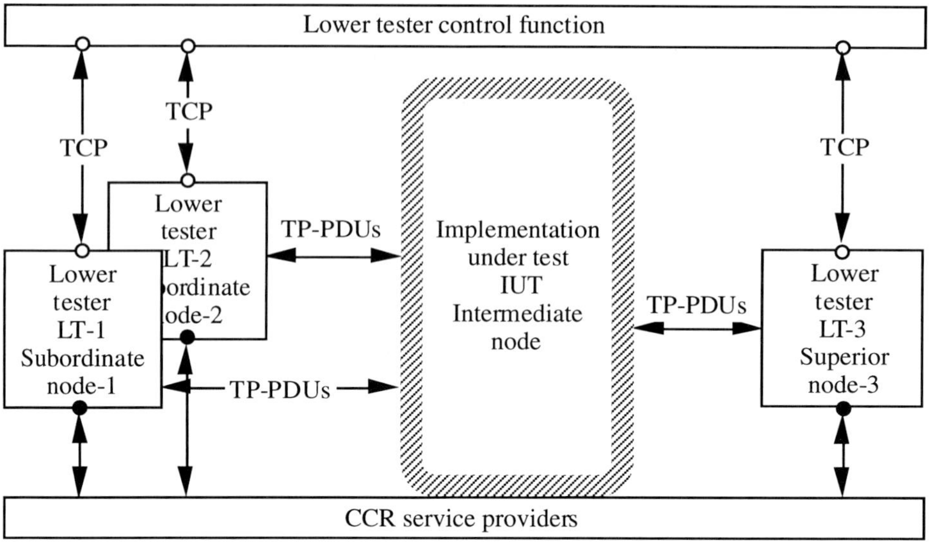

FIGURE 2.17. TP test configuration 2 – IUT in the role of an intermediate node

2.6 Abstract test suites

[ISO 9646-2] defines the requirements for the standardization of system independent abstract test suites (ATS) for OSI specifications. Moreover, it gives wide-ranging guidance on the development of test suites. Application specific testing with events not expressible in ASPs and PDUs, and the use of formal description techniques for creating or validating ATSs, are not addressed by 9646.

2.6.1 The production of an ATS

CTMF proposes the following steps for the ATS development process, which are not declared as binding, in particular with respect to their order:

- determination of the testable conformance requirements from the relevant OSI specifications, ICS proformas and profile requirements list, the latter being used only for profile testing.
- determination of test groups and, optionally, test group objectives (in the standardized sense) which cover these requirements,
- development of test purposes covering the requirements and test group objectives,
- choice of the testing context and an abstract test method: A test suite is always dedicated to a single test method; different test methods require dif-

ferent test suites; the choice of an abstract test method implies requirements on the lower and the upper testers (if included), and on the test coordination procedures;

- specification of the abstract test cases, including test steps, using a standardized test notation; specification of new profile specific test purposes and test cases, and standardization of the PSTS for profile testing; replacement of test purposes and test cases in base specifications according to the PSTS test case replacement list, which identifies the affected test cases;
- specification of the selection and parameterization criteria with respect to the ICS and IXIT proformas and, if appropriate, to the profile RL.

The ATS development process is usually linked with the development of related documents, such as TSS&TP, ICS and IXIT proformas, ATSs for other test methods, and the TMP specification. The ATS development steps listed above will be treated in more detail in the following sections.

2.6.2 ATS and ICS proforma

If static conformance requirements are incompletely or ambiguously specified in the OSI specification, the ICS proforma should be used to clarify them. If one does not exist, or does not clarify the open questions, the ATS specifier should write a defect report in order to have these problems solved by an amendment or revision of the respective OSI specification(s). Pending these clarifications, it can be assumed that

- capabilities mentioned in the specification but not included in the static conformance requirements are optional, but
- with respect to reception, everything without an explicit status should be implemented, and
- with respect to transmission, everything without an explicit status may be considered optional.

2.6.3 Test suite structure and test purposes

Test suites are hierarchically structured, as shown schematically in Figure 2.18. The main independent unit in a test suite is the *abstract test case*. A test suite consists of test cases, whose execution in a test campaign is not subject to any given order.

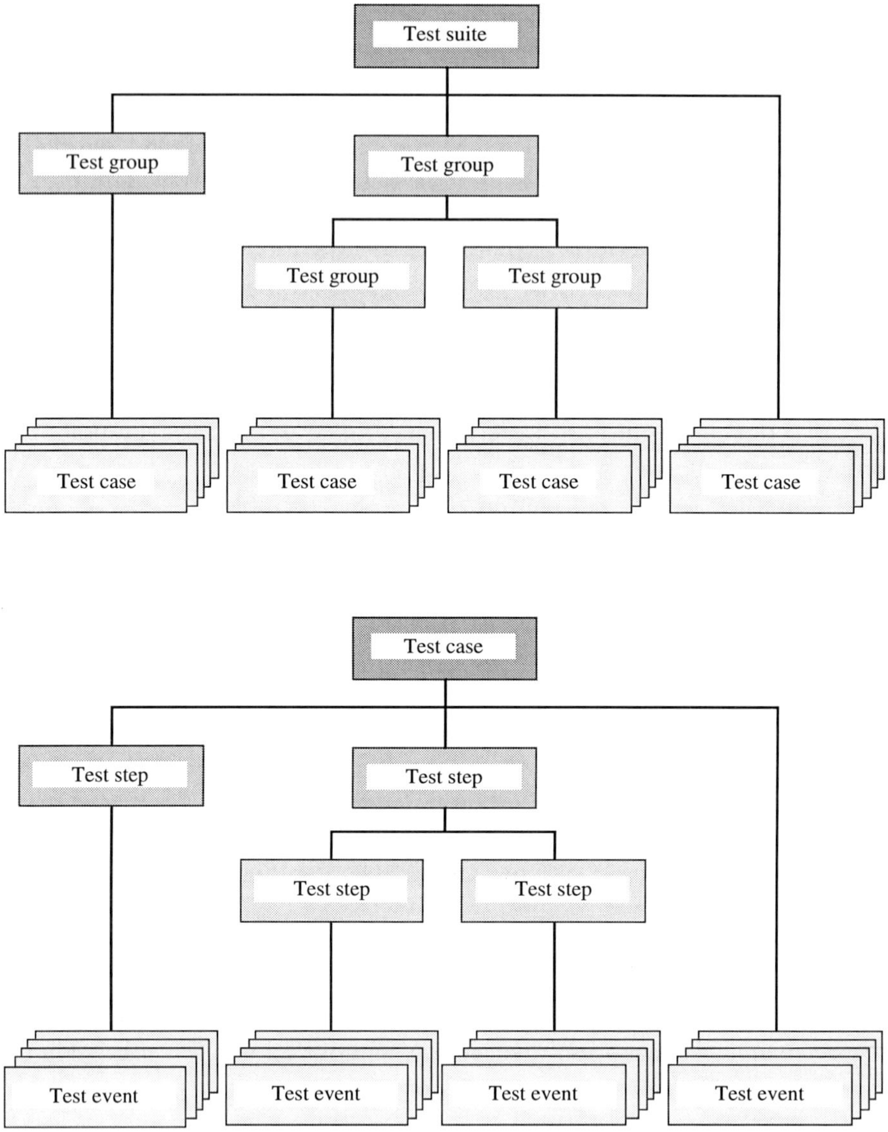

FIGURE 2.18. Test suite structure

Each test case normally has an individual *test purpose*, usually to demonstrate a certain behavior or the realization of some required capability. In order to facilitate both the systematic writing of an ATS, and its understanding by readers, test cases are logically grouped into *test groups*. Test groups themselves may be grouped further to form higher level groups, on as many levels as desired. As such groupings will usually be made based on a common, higher level goal of the associated test groups and test cases, a *test group objective* may be provided for each test group.

It is advisable in the production of the test purposes for a test suite to create the hierarchy of test group objectives in a top-down fashion, in order to ensure the systematic derivation of test purposes and to achieve good coverage of all protocol aspects and conformance requirements. Test purposes and test group objectives may be developed top-down, such that each consists of the concatenation of all objective and purpose aspects added along the path from the test suite through the nested test groups down to the test case.

Within test cases, the elementary units to be considered are *test events*. A test event is usually the exchange of a PDU or an ASP at a PCO. Of course, transfers of CMs at CPs (cf. Section 3.1) and timeouts are also possible test events. Each test case is made up of test events, which are normally combined under the two aspects of sequence and alternatives, thus forming what can be visualized as a tree (cf. chapter 3).

A test case may be modularized into test steps, and so may test steps themselves, to any depth of nesting. Modularization facilitates test case development, debugging and maintenance. It will often be economical to group commonly used test steps together in one or more *test step libraries*, such that they need not be specified repeatedly in different places. Test step libraries may be global for the test suite or dedicated to a particular test group.

Test purposes focus on one or more related conformance requirements. They are derived from whichever passage of the OSI specification defines them, e.g. state tables, PDU parameter or encoding descriptions or natural language behavior descriptions. Test purposes may be combined, in order to economize testing. These combinations are subject to restrictions. One test case may cover several fields of a PDU if this PDU is either received by the tester or sent with valid field contents. Invalid field value tests should not be combined with others. Combined test purposes should be listed in addition to the narrower test purposes grouped together to form them, with appropriate cross referencing provided. CTMF states that test purposes should be specified even for untestable conformance requirements contained in the protocol, which seems somewhat paradoxical.

The test purposes and their hierarchical grouping will be specified in a *test suite structure and test purposes* (*TSS&TP*) standard. As the conformance requirements of a protocol can usually be grouped into different categories, the actual form of the structure chosen depends largely on the personal experience and style of the TSS&TP writer. No matter what hierarchy is chosen, the test purposes for a test suite have to cover the following aspects of the protocol:

- capabilities, in particular mandatory and optional ones,

- behavior tests with the tester displaying only valid behavior,
- behavior tests with the tester displaying invalid behavior (syntactically or semantically invalid, or inopportune test events),
- PDUs sent to the IUT,
- PDUs received from the IUT,
- the relationship between PDUs sent and those received,
- protocol phases,
- protocol machine states,
- timing,
- encoding,
- individual parameters and parameter combinations.

This list is not claimed to be exhaustive, but rather a common core for protocol specific lists, which then extend this list to cover their more specific conditions. It is clear from this catalog that it takes considerable skill to mold a proper hierarchy from these different "dimensions," while at the same time avoiding overlaps and gaps in the coverage of the conformance requirements.

2.6.4 ATS specification

In an ATS, some general information on the test suite will precede the enumeration of the test cases:

- ATS name, date of origin and version number,
- names and version numbers of relevant protocol specifications (including an identification of the technical corrigenda taken into account), international standards or ITU-T recommendations of OSI service definitions, and of the specification of the test notation used,
- name of the test method for which the ATS applies,
- details of the coverage,
- description of the test suite structure,
- description of the test coordination procedures (or a reference to the TMP in the case of the coordinated test method),
- optionally, identification of the basic interconnection test cases,

- optionally, identification of the capability test cases,
- information for test realizers and test laboratories, cf. next section.

The use of TTCN (cf. Chapters 3 and 4) or any other test notation, standardized in the future, is recommended for the specification of test cases and test steps. Whenever the test suite specifier requires a means of expression which does not exist in the standardized test notation used, the necessary facilities must be pointed out in the ATS and proposed for inclusion in the test notation via a defect report or an amendment.

For each single or combined test purpose, either a test case is produced or reasons are stated explaining why this test purpose cannot be achieved under the chosen test method (e.g. it requires special features of the SUT).

Before we continue with the CTMF rules for ATSs, an explanation of some technical notions is in order: A *testing state* is a combination of the states of the SUT (including the IUT), the test system and the underlying provider which can occur during testing. It includes in particular the state of the protocols observed and controlled. Here CTMF seems to merge abstract states of protocol machines with real states of implementations. The *idle* testing state is the one in which no connection is established and which is independent of previously executed test cases – at least as far as subsequent observable behavior is concerned. A *stable* testing state is one which can be maintained between test cases, i.e. for a sufficiently long period of time and without explicit tester activities, e.g. a data phase basic state or the idle testing state.

Each test case is expected to end in one of a test suite specific set of stable states. This set shall include the idle testing state. Therefore, a test case is conceptually composed of a

- *test body*, the combination of test events which is responsible for achieving the test purpose and which starts in a particular *initial testing state*, a
- *test preamble*, which takes the test from each possible stable testing state, respectively, to the initial testing state of the test body, and a
- *test postamble*, which takes the test from any final state reached in the test body to one of the stable state.

The above definitions are founded on 9646-1. Part 2 of 9646 gives slightly different meanings to the terms preamble and postamble: for each stable testing state s, there exists a separate preamble, to take the entire testing configuration from s to the initial testing state. Similarly, part 2 speaks of several postambles for one test case.

If different stable testing states are possible between test cases, the test suite should specify that the identity of the last one attained be stored, so that the proper preamble can be chosen at the start of the next test case. These concepts permit an economical execution of test cases, avoiding the repeated traversal of disconnected idle states when test cases executed sequentially focus on the same advanced protocol phase – and all this without endangering the principle that test cases must be executable in any order.

Of course, the ultimate aim of the execution of a test case is to pass a judgement on what the test case revealed. What the test case can reveal is effectively a sequence of test events, the observed test outcome. For each test case, the test suite specifier must assign to each observable outcome a verdict, which is either "pass", "inconclusive" or "fail". In practice, this test verdict is often calculated from "preliminary verdicts" assigned to particular events in the test outcome. Test verdicts are treated in more detail in Subsection 2.3.5, as far as general principles are concerned, and in Section 4.6, as far as their assignment in TTCN is concerned.

2.6.5 Use and maintenance of an ATS

Additional information for the test realizer and test laboratory provided in the test suite includes details about

- the relationship between ICS entries and test case selection,
- a partial IXIT proforma, used to obtain the ATS parameters not provided by the ICS,
- the relationship between IXIT entries and test case selection,
- the order in which the test cases are to be listed in the PCTR, if different from the order in the ATS,
- restrictions on the order in which test cases may be executed,
- test cases or test groups which have to be realized in an MOT,
- test coordination procedures or a reference to a TMP,
- timing conditions.

Last, but not least, any ATS will require some maintenance, because its repeated use will reveal errors and gaps. The test suite specifiers will correct and extend the ATS according to the valid defect reports they receive. Moreover, possible changes of the OSI specifications will necessitate updates of the corresponding test suites.

2.7 Test realization

A real test system used to execute a test suite on IUTs is called a *means of testing* (MOT). In the sense of 9646, an MOT comprises the combination of equipment and procedures that can derive, select, parameterize and execute the test cases of a given ATS and can, in the course of their execution, produce a conformance log. *Test Realization* is the term used for the production of an MOT. In particular, it does not mean the performance of a test campaign, i.e. the actual execution of test cases, as the name might possibly suggest. Part 4 of CTMF states the requirements on test realization, which help ensure that the execution of test cases on an IUT by an MOT accurately reflects the behavior specified by the ATS. In keeping with the principle of implementation independence, the requirements on test realization are restricted to aspects which can be described in terms of abstract tester functions, cf. Subsection 2.5.5. Of course, test realization is based on the chosen ATS and the relevant OSI specifications.

2.7.1 Derivation of an ETS

Usually, an ATS is not directly executable, even if the required parameter values and selection information are available as an additional input. Rather, test realizers must convert the abstract test cases of the ATS into a form which can be executed by a test system. The converted form of the ATS is called an *executable test suite* (ETS). However, what is actually executed in a test campaign will also depend both on the selection of applicable test cases and on the parameter values to which the test suite parameters are bound. Selection and parameter assignment are performed on the basis of the ICS and IXIT answers. Although CTMF slightly falters between the equivalent views of the selection of applicable test cases and of the deselection of the complement set, let us simply speak of selection. Also, although what is really meant is the elimination of parameter names by replacing them with values, let us call this process *parameterization*, in agreement with the standard. CTMF recognizes that different orders of derivation, selection and parameterization are imaginable, as shown in Figure 2.19. The resulting forms of the test suite, apart from the ATS and the ETS, whether tangibly existing or not, are called selected abstract test suite (SATS), selected executable test suite (SETS), parameterized abstract test suite (PATS) and parameterized executable test suite (PETS). Only the ATS must be a tangible entity.

Requirements on the selection and parameterization processes, being closely linked with the interaction between the client and the test laboratory, are given in the following section. As to the derivation process, 9646 states the following requirements:

- There is a one-to-one correspondence between selected abstract and executable test cases.
- The executable test case derived from an abstract test case must correspond to the latter in all necessary respects: behavior, test purpose, test verdicts, analytical power (e.g. no additional PDU field checking), relation to ICS/IXIT entries, selectability, independence from other test cases, possibility of arbitrary ordering, etc.
- Apart from the ICS and IXIT determined selection process described above, the derivation process may include, although this cannot be strongly recommended, some MOT specific selection.
- Derivation must be independent from and compatible with the selection and parameterization processes, such that all paths through Figure 2.19 yield the same result.

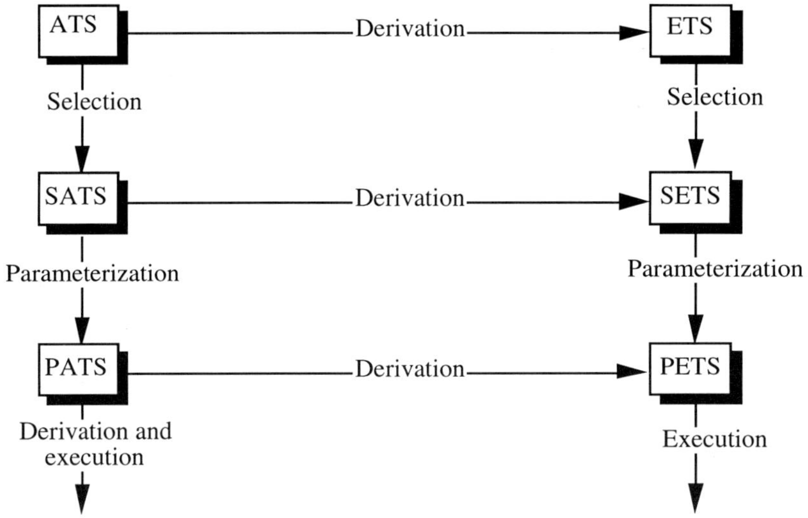

FIGURE 2.19. Forms of test suites

2.7.2 Means of testing

The means of testing are provided with respect to a single standardized ATS and, of course, only for the latest version of the ATS with the highest status (IS rather than DIS rather than DP or CD, latest version, if several exist). The MOT shall

- realize the lower testers and, if appropriate, the lower tester control function,
- specify or realize the upper testers, the test coordination procedures and the TMP, each depending on the abstract test method chosen in the ATS,
- either contain derived executable test cases or permit their derivation, both, of course, with the correct semantics,
- provide a means for selection and parameterization,
- provide a means for executing the selected, parameterized, executable test cases of the PETS,
- provide a means of generating a conformance log.

As described in Section 2.5, the lower testers, being the peer entities to the IUT, have the main responsibility for running the executable test cases, for managing test coordination procedures, and for producing preliminary test verdicts and passing them to the lower tester control function. The lower testers are implemented within the independent test system, where they operate on top of the underlying service. There are two major alternatives for the realization of the lower testers:

- In the *encoder/decoder* technique, the LTs merely encode and decode the ASPs and PDUs as required by each test case, potentially without any knowledge of the protocol mechanisms apart from those contained in the active test case. The advantage of this technique is that the avoidance of a full implementation of the protocol makes the implementation job simpler and faster. The disadvantage is that the executable test cases must contain all the protocol information, possibly redundantly and with a considerable demand of storage space.
- In the *enhanced implementation* (sometimes also called *reference implementation*) technique, a protocol implementation is extended by the ability to generate erroneous PDUs and by a test driver module, which controls the implementation based on the given test cases. Certainly more demanding to realize, this variant permits a reduced version of the executable test cases, to be produced, as the protocol conforming behavior of the tester is realized automatically and need not be written down in detail – only as far as choices between alternatives are concerned.

As mentioned in Section 2.5, upper testers can exist in various forms, e.g. in the form of dedicated software implemented in the SUT above the IUT, as a human operator, or, in the remote case, realized by the existing protocol layers above the IUT. In either case, the UT realization must be able to cope with the ASPs as they are realized in the SUT.

The realization of the test coordination procedures depends on the realization of the LTs and UTs. Choices are possible, such as whether to work with or without synchronization, whether to use the given service provider or to deploy other communication channels, etc. For example, a human UT can be prompted by the LT, a "scenario interpreter" can read its instructions from a file, upper and lower testers realized by software can synchronize by means of a TMP.

2.7.3 Documentation and maintenance

We can distinguish between two kinds of documentation pertaining to test realization: documentation on the MOT, which we will deal with first, and documentation produced by the MOT in the course of the test campaign, which we will treat subsequently.

MOT documentation has to include the MOT identification information (name, date, version, ETS identification, if any), references to protocol and ATS specification documents, including corrigenda, an MOT description (details about the realization of the abstract testing functions and their coordination, about selection and parameterization), a list of test cases to be excluded due to limitations of the MOT, information on the procedures for test execution and on the interpretation of the conformance logs produced. Also required are statements of conformance to the ATS concerned and of compliance with 9646-4.

When an executable test suite is run against an IUT, the MOT has to keep a human readable conformance log which permits verfication of the test verdicts assigned. This implies that the MOT must be able to produce a paper version of the log that can be understood by a human reader. The test realization part of 9646 specifies in detail mandatory information to be included in the conformance log, including

- substantial information such as the order and contents of the ASPs and PDUs sent and received by the lower and upper testers, the order and content of CMs sent and received by the LTCF, an indication of the PCOs and CPs (see 3.4.7) used, test case identifiers, preliminary and final verdicts, and indications of test case errors or abnormal test case terminations;

- administrative information such as proper identification of the conformance log, start date of PETS execution, identification of the MOT and, if applicable, ETS as well as test case identification, plus start and end time, for each test case.

In the conformance log, the terminology of the relevant protocol or transfer syntax specification, as well as, wherever not in conflict with the latter, that of the ATS standard, has to be used to record events.

Maintenance, too, has two aspects regarding MOTs. On the one hand, an MOT itself is maintained; on the other hand, it helps to contribute to the maintenance of protocols and test suites. Problem reports with respect to abstract or executable test cases, usually originating from test laboratories, have to be analyzed to determine whether the MOT is not conforming with the ATS or not complying with 9646, or whether there are problems with the ATS or even the OSI specifications. The test realizer is expected to make corrections or file defect reports, depending on the source of the problem.

On of the most important matters concerning the success of 9646 is the correction of ATSs, coupled with proper reporting of the fixes. Therefore, it is essential that MOT providers and the test laboratories who detect errors do not view their individual fixes as "selling points," but report them to the standardization groups. Otherwise, different fixes will cause different test suites to be used for the same protocol and the same test method.

2.8 The conformance assessment process

As we have shown, testing an IUT involves far more activities than the mere execution of the ATS. The completion and initial analysis of pretest documents, the adaptation of the ATS to the particular capabilities and parameter values claimed to be implemented, its physical realization, as well as agreements on test coordination procedures, are activities scheduled prior to the actual running of a test suite. The actual execution of the test suite is called the *test campaign*. During and after the test campaign, posttest documents will be produced. The entirety of all these activities, which are necessary to validate and document the conformance of a protocol implementation to the relevant OSI specifications, is called the *conformance assessment process*. This process spans three distinct phases:

- preparation for testing,
- test operations, and
- test report production.

The preparation for testing involves the production of SCS, ICS and IXIT as necessitated by the number of specifications involved, the choice of abstract test method and ATS, the preparation of the system under test (SUT) and the means of testing (MOT).

The test operations involve a static conformance review of the ICS, test selection and application of parameter values, and one or more test campaigns (one for each relevant ATS).

The test report production process provides the clients with documentation they can use for further steps, such as obtaining a certificate.

Figure 2.20 relates the various activities and documents relevant to the conformance assessment process. Part 5 of [ISO 9646] describes the requirements on the test laboratory and the client for conducting the conformance assessment process.

2.8.1 Test laboratories and test clients

The test laboratory is responsible for conducting the conformance assessment of IUTs. According to their role with respect to the acquisition process of the IUT, test laboratories can be classified as one of the following:

- *First-party* test laboratories are operated by the supplier (implementer, vendor) of the IUT tested;
- *second-party* test laboratories are affiliated to the procurer or user of the IUT;
- *third-party* test laboratories are independent from either of the two.

Clients will frequently be implementers or vendors of protocol implementations who want to have their own products tested and possibly certified. It is also imaginable that important procurers of IUTs might be interested in having the quality of these IUTs assessed, if the supplier has not already had recognized conformance testing done.

2.8.2 Preparations for testing

Table 2.15 shows the steps undertaken in this phase. The client has to ensure that the IUT is testable by at least one abstract test method. In multi-protocol implementations this applies to each protocol in the IUT. Of course, the appropriate facilities for test control and observation, and possibly test management, have to be provided, in accordance with the chosen test method, as described in Section 2.5. During preparations for testing, and apart from the core documents (ICS, IXIT and SCS), the test laboratory and the client exchange information in order to reach agreement on the definition of the IUT and the choice of abstract test methods and abstract test suites. In addition to the formal declarations of adher ence to standards (compliance with 9646-5, conformance with ATS standards,etc.), the test laboratory must provide the following information:

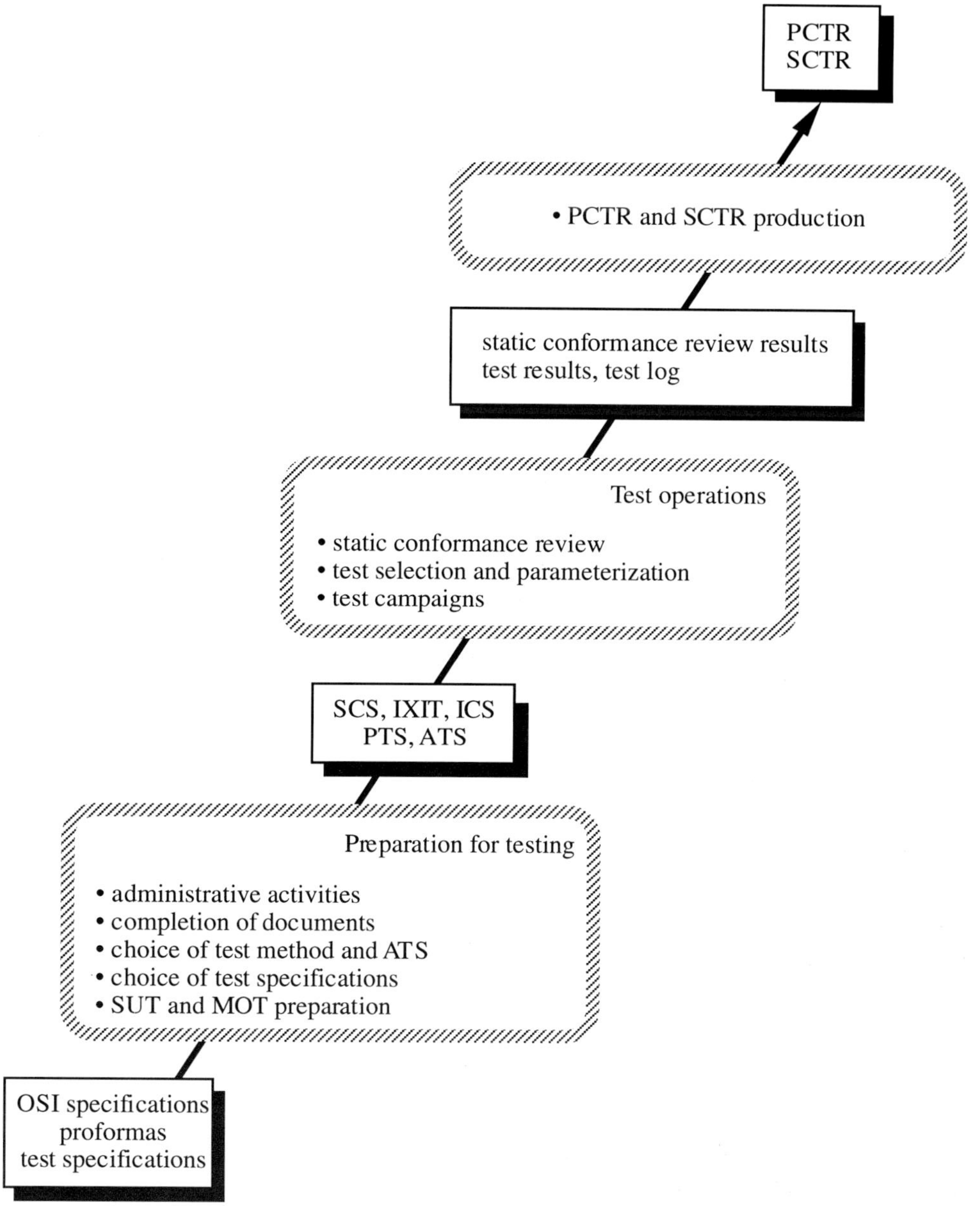

FIGURE 2.20. An overview of the conformance assessment process

- which ATMs are supported for which protocols;
- instructions for providing the SCS, ICSs and IXITs, in particular for completing the respective proformas;

- whether comprehensive testing service is offered (9646-4 does not elaborate on what this means);
- limitations, if any, of the LT's capabilities;
- details on the UT's operations, the TCPs, and the test laboratory procedures for running test cases during the test campaign, as far as they have beings on the SUT or its operator;
- a list of the documents used or produced by the test laboratory during the conformance assessment process.

TABLE 2.15. Preparation for testing

Input	Activities	Output
	• request for testing service, by client	
	• agreement on contractual terms and conditions of testing	
	• provision of documents and other information by the test laboratory	e.g. ICS, SCS proformas
	• provision of necessary information by the client	
ICS, SCS proforma	• completion of proforma by the client	ICS, SCS
ICS, SCS, SUT configuration and testability descriptions	• agreement on test methods and ATSs	test method, ATSs
TMP implementation statement proforma	• completion of TMP implementation statement by the client, if the coordinated test method has been chosen	TMP implementation statement
ATSs, MOT	• test laboratory determines MOT and produces IXIT proformas	IXIT proformas
IXIT proformas	• completion of IXIT proformas	IXITs
IXITs, SUT	• preparation of the SUT by the client	SUT ready to be tested
IXIT, MOT	• preparation of the MOT by the test laboratory	MOT ready for testing

The test laboratory may provide the client with additional information and help, such as information on the testing services offered, time estimates, accreditation status, assistance in the realization of the UT, etc.

Apart from the formal statements required, such as compliance with 9646-5 and the claim of testability using specific ATMs and ATSs, the client must provide the following information: which part of the SUT makes up the IUT, which protocols are to be tested, and which TCPs are supported.

Additionally, the client may provide the test laboratory with information such as particular physical requirements, if any, of the SUT, or a contact person during conformance assessment. This information is used by the test laboratory and the client, to agree on the definition of the IUT and on the ATM(s) and ATS(s) to be used. If this agreement cannot be reached, the conformance assessment process is terminated via a negotiated exit. If agreement can be reached, this fact and the agreed details are recorded in the SCTR (cf. 2.8.4).

2.8.3 *Testing operations*

The main steps undertaken in the testing operations phase, and their interactions, are listed in Table 2.16. The test campaign includes, to name only the most important items, the assignment of verdicts, production of the conformance log and negotiated exits.

TABLE 2.16. Testing operations

Input	Activities	Output
ICS, SCS, OSI specification	static conformance review	SCR result
ICS, IXIT, ATS	test selection	SATS
ICS, IXIT, SATS	test parameterization	PETS
PETS	verification of TCPs	PETS
PETS	test campaign	test result

STATIC CONFORMANCE REVIEW

During the static conformance review, the test laboratory checks the ICSs for consistency with the static conformance requirements of the relevant OSI specifications. This analysis is done based on the following questions:

- Have all mandatory capabilities been indicated as implemented in the support column?
- Are the answers for all optional and conditional capabilities consistent?

– Are all additional static conformance requirements stated in the ICS proformas fulfilled?.

Moreover, the test laboratory has to check the information provided by the client in the ICS, SCS and IXIT documents for consistency.

TEST SELECTION

Based on the information provided in the ICS and IXIT, the test laboratory selects a suitable subset of test cases from the relevant abstract test suites. The result of the test case selection is presented in the "selected" column of the protocol conformance test report (PCTR, see subSubsection 2.8.4). Test cases which have been deselected for operational reasons, like ETS error, no support in MOT, etc., are marked as "not run" in the "Run" column of the PCTR. The resulting selected abstract test suite (SATS) includes capability and dynamic behavior test cases for mandatory capabilities, as well as for optional and conditional capabilities which have been indicated as implemented in the ICSs. Additionally, the test laboratory may select basic interconnection tests, if this service is requested by the client and if the BITs have been specified in the ATS.

TEST PARAMETERIZATION

After test selection, the test laboratory performs the test parameterization. During this phase, parameters are assigned values as specified in the ICS and IXITs. In addition, the transformation of parameterized selected abstract test cases into executable test cases is completed. The result of this step is the parameterized executable test suite (PETS). Typical values assigned during parameterization include network addresses, connection endpoint identifiers, counter variables or timer values, etc. Of course, it is possible to change the order of test selection and test parameterization, as indicated in Figure 2.19.

VERIFICATION OF TCPS

Every ATS contains explicit or at least implicit specification statements about test coordination procedures. The test laboratory has to verify that the MOT and the SUT are able to use these TCPs. In the case of the remote or distributed test method, it is sufficient for the client and test laboratory to check the TCPs informally. In the case of the coordinated test method, the test laboratory has to check the implementation of the test management protocol in the SUT. This is done using the TMP implementation statement provided by the client in order to select the appropriate TMP test cases and to run them against the upper tester. If the local test method is used, then the test laboratory verifies the TCPs together with the MOT.

TEST CAMPAIGN

A test campaign comprises the execution of the paramaterized executable test suite and the production of the conformance log. To accomplish this, the tasks of the test laboratory also include the allocation of test staff and the MOT. Similarly, the client has to provide the SUT and an SUT operator. In addition to running the SUT during the test campaign, the SUT operator is also responsible for any changes to the IUT or the IUT environment, as required and negotiated with the test laboratory. In principle, these modifications are only allowed in exceptional cases. They must not invalidate the test results of completed test cases. Further, all modifications which have been made by the client or the test laboratory during the test campaign have to be carefully documented, so that a decision can be made as to whether or not the test campaign should be repeated, continued or aborted. If basic interconnection testing is requested, the test laboratory must run the relevant test cases first. When completed, the client has the opportunity to review the results and to request that the test laboratory allow negotiated exit. If the BITs are successful, the test laboratory continues with the capability and behavior tests.

ASSIGNMENT OF VERDICTS

The test laboratory has to assign exactly one test verdict per executed test case. In the case of abnormal test case termination or executable test case errors, the test laboratory has to repeat the related test cases. If the same undesired test result is achieved, the test laboratory has to note this fact by entering "not run," accompanied by an appropriate explanation of the problem, in the PCTR.

In our opinion, an executable test case error must lead to an investigation of the problem: is it an error in the abstract test case, in the executable test case derivation, or in its execution? Unfortunately, the standard does not require this.

For all those test cases which have a recognized abstract test case error, the test laboratory has to indicate "not run" and the reason for it in the PCTR. It would of course make sense to deselect such tests before actual testing begins.

All test cases with the test verdict "inconclusive" have to be repeated at least once by the test laboratory. If a pass or fail verdict is achieved during the repetitions, then this verdict has to be recorded in the PCTR. However, if the inconclusive verdict is reproduced during the repetitions, then "inconclusive" has to be recorded in the PCTR.

For test cases which end with a fail verdict, the test laboratory has to determine whether the test verdict has been assigned to an unidentified test event, i.e. a test event that is not among a specified list of ASP and PDU types. If this is not the case, then the test verdict fail will be recorded. If this is the case, the test laboratory has to verify that an abstract test case error has not caused valid behavior to be accidentally subsumed under the unidentified test event.

Before the test laboratory starts the production of the test report, it informs the client about all test cases which produced a fail verdict. Clients may request repetition of failed test cases if the test laboratory is unable to demonstrate the existence of an error in the IUT.

CONFORMANCE LOG

At the request of the client, the test laboratory has to produce a conformance log, which documents the outcome of all executed test cases, including repetitions.

NEGOTIATED EXIT

Negotiated exits (NE) are points in time when the client and the test laboratory jointly come to the conclusion, based on the most recent test results, that a continuation of the test campaign is useless. An SCTR or PCTR is not produced by the test laboratory after the occurrence of a negotiated exit. Instead, the test laboratory produces an informal test report and, if requested by the client, a conformance log of the test cases executed. A test campaign cannot be continued after a negotiated exit. Only a completely new test campaign can be started.

2.8.4 Test report production

The conformance assessment process ends with the production of a system conformance test report, plus a protocol conformance test report for each tested protocol, provided that no negotiated exit occurred.

SYSTEM CONFORMANCE TEST REPORT

At the end of the conformance assessment process, the test laboratory produces an SCTR, which contains a summary of the achieved test results and additional administrative information. Separate SCTRs are required if an SUT is tested for more than one profile. An SCTR uses the schema presented in Table 2.17, where some explanations are given in italics. Note again, that the information indicated by "• • •" is to be provided by the test laboratory and that the information indicated by "– – –" is to be provided by the client.

TABLE 2.17. System conformance test report proforma

SYSTEM CONFORMANCE TEST REPORT FOR	• • •	*Name of the SUT*
1. IDENTIFICATION SUMMARY		
This section contains administrative information for identification of the SCTR		
1.1 SYSTEM CONFORMANCE TEST REPORT		
SCTR number	• • •	
SCTR date	• • •	*Date of issue*
Test laboratory manager	• • •	
Signature	• • •	
1.2 TEST LABORATORY		
This section has the same structure as the corresponding section in table 2.9.		
1.3 TEST CLIENT		
This section has the same structure as the corresponding section in table 2.9.		
1.4 SUT		
This section contains information about the SUT.		
Name	• • •	
Version	• • •	
Supplier	• • •	
Period of testing	• • •	
Date of receipt of SUT or		
Location of SUT for testing	• • •	
SCS identifier	• • •	*Reference to attached documents*
1.5 PROFILE		
This section applies only to profile SUTs		
Profile identification	• • •	
Profile version	• • •	
PTS summary reference	• • •	
PSTS references	• • •	
Profile ICS	• • •	
Profile IXIT	• • •	
1.6 NATURE OF CONFORMANCE TESTING		
The purpose of Conformance Testing is to increase the probability that different implementations can interwork. However, the complexity of OSI protocols makes exhaustive testing impractical on both technical and economic grounds. Furthermore, there is no guarantee that an SUT which has passed all the relevant tests conforms to a specification.		

TABLE 2.17, continued. System conformance test report proforma

Neither is there any guarantee that such SUT will interwork with other real open systems. Rather, the passing of the tests gives confidence that the SUT has the stated capabilities and that its behavior conforms consistently in representative instances of communication.

1.7 LIMITS AND RESERVATIONS

Additional information relevant to the technical contents or further use of the test report, or to the rights and obligations of the test laboratory and the client, may be given here. Such information may include restrictions on the publication of the report. • • •

The test results presented in this test report apply only to the particular SUT and component IUTs declared in sections 1.4 and 1.8 of this SCTR, for the functionality described in the referenced SCS and in the ICS(s) referenced in each PCTR, as presented for test in the period declared in section 1.4 and configured as declared in the relevant IXIT(s) referenced in each PCTR. This SCTR may not be reproduced except in full together with its SCS.

1.8 RECORD OF AGREEMENT

A definition of what parts of the SUT were considered to be the IUT during testing, and of the abstract test methods and abstract test suites that were used.

IUT definition reference	Component	ATS reference	ATM(s)
• • •	• • •	• • •	• • •
• • •	• • •	• • •	• • •

The above table gives guidance to the nature of the information which may appear in the record of agreement. Here the IUT definition reference column means the full reference of the IUT to be tested by referenced ATS. A component identifies a protocol and/or information object which is to be tested. The ATMs column contains a list of ATMs used by the ATSs and also indicates if multi-party methods are being used.

1.9 COMMENTS

Additional comments may be given by either the client or test laboratory on any of the contents of the SCTR, for example, to note disagreement between the two parties. • • • /– – –

2. SYSTEM REPORT SUMMARY

For each component (i.e. protocol or set of protocols or information object) tested, a summary of the testing and conformance status of the implementation of that component is required, using clauses of the format shown below.

2.n TESTING SUMMARY FOR	• • •	*Component name*
Accreditation status	• • •	*Reference*
Accreditation reference	• • •	*Reference*
Implementation identifier	• • •	*Name and version number*
IUT Definition reference	• • •	*Reference from section 1.8*
Protocol specifications	• • •	*References*
Information object specifications	• • •	*References*
ICSs	• • •	*References*

TABLE 2.17, continued. System conformance test report proforma

IXITs	• • •	*References*
PCTR number	• • •	*Reference*
PCTR date	• • •	*Date of PCTR*
ATS specification(s)	• • •	*References*
Abstract test method(s)	• • •	*Identification of methods*
Means of testing identifier	• • •	*Name and version number*
Conformance status		
Static conformance errors?	• • •	*Yes/No*
Dynamic conformance Errors?	• • •	*Yes/No*
Test cases run	• • •	*Number*
Passed	• • •	*Number*
Failed	• • •	*Number*
Inconclusive	• • •	*Number*
Observations (optional)		
If the SUT is not statically and dynamically conforming to this protocol, an additional summary may be given on aspects of non-conformance. Any difficulties encountered may be reported here.	• • •	

PROTOCOL CONFORMANCE TEST REPORT

At the end of the conformance assessment process, the test laboratory produces a protocol conformance test report for each tested component, that is for each protocol or information object. Besides administrative and technical information, the PCTR contains both a summary and an individual presentation of all achieved test results. The schema for PCTRs is shown in Table 2.18 with some explanations in italics.

TABLE 2.18. Protocol conformance test report proforma

PROTOCOL CONFORMANCE TEST REPORT FOR	• • •	*Component description*
1. IDENTIFICATION SUMMARY		
1.1 PROTOCOL CONFORMANCE TEST REPORT		
PCTR number	• • •	
PCTR date	• • •	*Date of issue*
Test laboratory	• • •	*Identification*

TABLE 2.18, continued. Protocol conformance test report proforma

Accreditation status	• • •	*Reference*
Accreditation reference	• • •	*Reference*
Technical authority	• • •	*Name*
Job title	• • •	*Name*
Signature	• • •	
Test laboratory manager	• • •	*Name*
Signature	• • •	
1.2 IUT		
Name	• • •	
Version	• • •	
Protocol specification(s)	• • •	*References*
Information object specifications	• • •	*References*
ICS(s)	• • •	*References to attached documents*
Profile RL	• • •	*References*
Profile specific ICS	• • •	*References*
1.3 TESTING ENVIRONMENT		
IXIT(s)	• • •	*References to attached documents*
Profile XRL	• • •	*References*
Profile specific IXIT	• • •	*References*
ATS specification	• • •	*Reference*
Abstract test method(s)	• • •	*Method identification*
Means of testing identification	• • •	*Names and versions of MOT*
Period of testing	• • •	
Conformance log reference(s)	• • •	*Information required to obtain conformance logs*
Retention date for log reference(s)	• • •	
1.4 LIMITS AND RESERVATIONS		

Additional information relevant to the technical contents or further use of the test report, or to the rights and obligations of the test laboratory and the client, may be given here. Such information may include restrictions on the publication of the report. • • • / – – –

The order of test cases listed in section 6 of this PCTR corresponds to the ordering of test cases defined in the ATS referenced in section 1.3. This does not indicate that the test cases were executed in this order. The test results presented in this test report apply only to the particular IUT declared in section 1.2 of this PCTR, for functionality described in the relevant ICS(s) attached to this PCTR, as presented for test in the period declared in section 1.3 and configured as declared in the relevant IXIT(s) attached to this PCTR. This PCTR

TABLE 2.18, continued. Protocol conformance test report proforma

shall not be reproduced except in full together with its attached ICS(s) and IXIT(s).

1.5 COMMENTS

Additional comments may be given by either the client or test laboratory on any of the contents of the PCTR, for example, to note disagreement between the two parties. • • • / – – –

2. IUT CONFORMANCE STATUS

This IUT has/has not been shown by conformance assessment to be non-conforming to the referenced base specification(s). *Strike the inappropriate words, "has" or "has not," in this sentence.*

3. STATIC CONFORMANCE SUMMARY

The ICSs for this IUT is/is not consistent with the static conformance requirements in the referenced base specifications. *Strike the inappropriate words, "is" or "is not," in this sentence.*

4. DYNAMIC CONFORMANCE SUMMARY

The test campaign did/did not reveal errors in the IUT.

Strike the inappropriate words, "did" or "did not," in this sentence. In addition, a summary of the results of test groups may be given. The detailed results of testing are provided in the table of section 6. This summary may, for example, give totals for the number of passes, fails and inconclusives in each test group, and also allows the test laboratory to make observations on those results, such as "All the tests concerned with segmented data transfer failed."

5. STATIC CONFORMANCE REVIEW REPORT

If section 3 indicates non-conformance, this section itemises the mismatches between the ICSs and the static conformance requirements of the referenced base specification(s). • • •

6. TEST CAMPAIGN REPORT

ATS Reference	Selected?	Run?	Verdict	Observations
	Yes/No	*Yes/No*	*P/ F/ I*	
• • •	• • •	• • •	• • •	• • •

7. OBSERVATIONS

Additional information relevant to the technical content of the PCTR may be given here.

• • •

Section 6 of the PCTR is the most important part of the test report, and will be explained in some more detail. The proforma for the test campaign section represents the minimal information required for the documentation of the test results. Additional columns may be added for providing further information, for example the ATS to ETS transformation or conformance log references. Test results are shown in the results table in the same order as they are defined in the

PSTS for profile testing or in the ATS for protocol and information object testing.

The first column references the abstract test case, usually by giving the test case identifier. In the second column it is indicated whether the test case was selected for testing according to the information provided in the ICSs and IXITs for the IUT. The reasons for deselecting test cases are given in the "observations" column. This column is also used to indicate the reasons for incomplete test case executions. The achieved test results are presented in the fourth column. Finally, the last column may also contain references to Section 7, for example references to explanations of why particular test cases produced a fail or inconclusive verdict.

CHAPTER 3. TTCN - THE STATIC PARTS

The next two chapters are based on the first IS version of TTCN (*Tree and Tabular Combined Notation* [ISO 9646-3]), published in 1992, plus preliminary versions of the two amendments specifying TTCN extensions, [ISO 9646-3-1 and 2]. The importance and extent of the amendments prompted us to take account of them in this book, even at the risk of having slight deviations from their final versions, which are due in 1994.

A "grandfather clause" in TTCN permits the use of the DIS version of TTCN in all abstract test suites that reached DIS status by 1991 or became CCITT Recommendations by 1992. It was included in order to avoid invalidating the work invested in test suites based on earlier versions of TTCN (e.g. for X.25), and to respect their advanced standardization status. Differences between the DIS and the IS versions of TTCN are listed in an annex to [ISO 9646-3]. We will not discuss them in this book.

An introductory remark in the standard clearly describes TTCN as an informal test notation. Opinions on the degree of formality achieved in TTCN vary among experts. We will try to give our readers an idea of the respective strengths and weaknesses of TTCN. Obviously, a considerable attempt to formalize the notation was made, and the disclaimer might be meant to pacify readers who discover passages where this attempt was not carried to completion.

3.1 Conventional TTCN and concurrent TTCN

In the 1992 version of [ISO 9646-3], a test case written in TTCN was abstractly defined as a single process which may be able to interact with its environment at one or two PCOs, even if the two are remote from one another. A test case description for a test method with upper and lower testers describes the possible sequences of PDU or ASP events at the PCOs. A PCO is an interaction point for communication between either the lower tester and the underlying service provider, or the upper tester and the IUT. It provides two unbounded FIFO queues for PDUs and ASPs, cf. Figure 3.1. Event sequences at different PCOs are interleaved in a prescribed manner. Test coordination procedures, which are not fully specified in the standard, have to ensure that the proper global order of events is achieved. Earlier attempts to include "parallel trees" had been deferred, because they could not be brought to maturity before the envisaged publication date of the TTCN standard. However, work on explicit concurrency in TTCN proceeded, and will be included in the first amendment to TTCN [ISO 9646-3-1].

The main idea behind concurrent TTCN is to provide for simultaneously executable abstract test components, each of which processes its own behavior specification. These *test components* may interact via PCOs with the underlying service provider or the IUT, and they may communicate with one another via asynchronous message transfer, realized at *coordination points* (*CP*). Each CP permits communication between exactly two test components and provides two unbounded FIFO queues for *coordination messages* (*CM*), one queue in either direction, as shown in Figure 3.1. CMs may be of any type that can be declared with the existing methods of type definition. Send actions at CPs, used to put CMs into the queue, are expressed like send events at PCOs. Receive actions at CPs, used to get CMs from the queue, are expressed like receive events at PCOs.

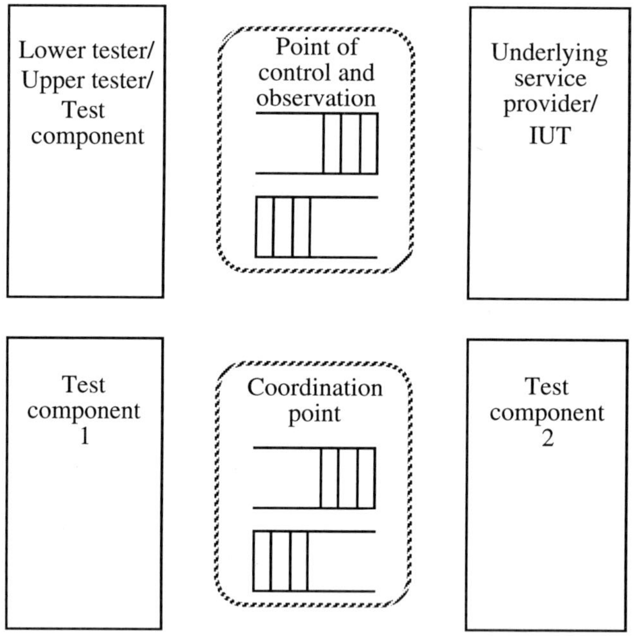

FIGURE 3.1. The structure of PCOs and CPs

Test components may use an arbitrary but statically limited number of PCOs and CPs. They can interact with the IUT or an underlying provider, both via PCOs. They can communicate with other test components, via CPs. There may be more than one CP between two test components, although it is not obvious when this would be needed. Conversely, even a test component without any

PCO or CP can be used, although there is not much it can do for the other components.

There is also a certain amount of implicit interaction between the test components, which does not use CMs or CPs, namely by the use of DONE checks and test verdicts. We will deal with these below.

The assignment of specific PCOs and CPs to test components defines the *test component configuration*, which is the actual "abstract architecture" for any given test case. The fictitious configuration example shown in Figure 3.2 gives an idea of the wide range of connectivity options. Test component configurations may vary from test case to test case. Different test cases may use different test components, but they may also use the same components, even in the same configuration.

In each test component configuration, exactly one test component is a *main test component* (MTC); the others are *parallel test components* (PTC). These roles are fixed independently of the configuration, when each test component is declared. So is the test component's name, and the number of PCOs and of CPs it supports, but not the identity of these interaction points.

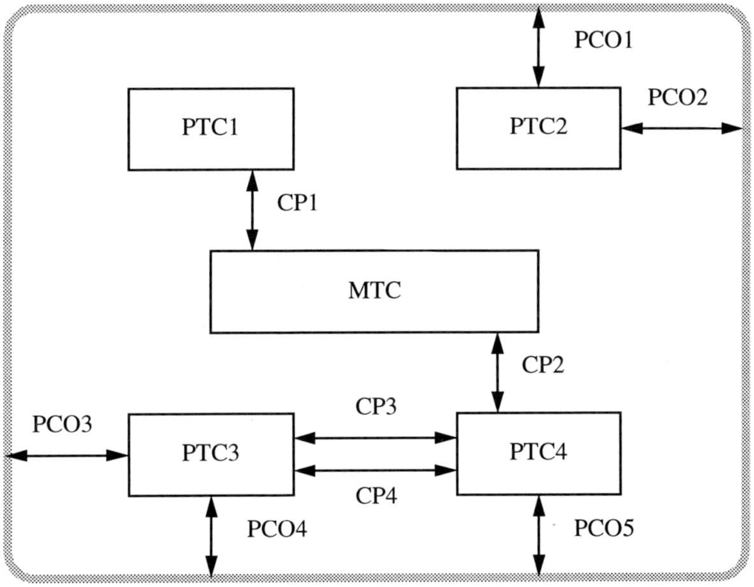

FIGURE 3.2. A test component configuration

It is left to the discretion of the test suite writers to choose how to declare test components. They may define several test components with the same roles and numbers of PCOs and CPs, in order to give these components self-explanatory names, thereby indicating the particular tasks they perform. Or, they may prefer to have a minimal number of test components, which serve as formal placeholders for each required combination of the role and the numbers of PCOs and CPs. In different test cases, the same test component may be assigned a variety of behaviors.

Not all test components in the configuration declared for a test case become active when the test case is started; in fact, only the MTC is started automatically. The MTC can start PTCs by means of explicit CREATE statements in its behavior description. The MTC is also the only component which can assign a final verdict and terminate the entire test case – figuratively speaking, it is the first to come and the last to go. All PTCs are expected to have terminated before the MTC does, and the latter may use DONE statements to find out whether they really have. This is the first method of implicit communication between test components. Preliminary verdicts affect the other kind of implicit communication: a verdict assigned in a PTC influences the MTCs global result variable in the same way as a preliminary verdict in the non-concurrent case.

More details of concurrent and non-concurrent TTCN will be described in various contexts throughout this chapter and the next.

3.2 TTCN.GR and TTCN.MP

TTCN can be written in either of two notations, one a graphically enhanced form for the human reader, *TTCN.GR*, and the other a machine processable form, *TTCN.MP*.

In the GR form, a test suite is a sequence of tables subdivided into boxes. In the MP form, a test suite is a single string of characters from the ISO 10646 alphabet [ISO 10646]. This alphabet, which is also called UCS (Universal Coded Character Set), comardized previously on a national or international level. The following characters will suffice for most ATSs:

- the letters from a to z and from A to Z
- the characters 0 1 2 3 4 5 6 7 8 9 , . ; : _ + - * / = > < ' \ ? ! ( ) [ ] $
- the blank symbol.

Other characters may only appear in what is called *free text*, cf. Section 3.2.4. Entries in GR tables are formed from the same characters, usually without the $-symbol, which is used to denote BNF keywords in TTCN.MP. Note that the notion of keywords has two different meanings in TTCN. BNF keywords start with a $ symbol; they delimit substrings corresponding to TTCN.GR tables, subtables and table entries, and facilitate TTCN.MP parsing. TTCN keywords like

GOTO and OTHERWISE, on the other hand, have important semantic functions; together with the TTCN predefined identifiers and the ASN.1 reserved words they are reserved words in TTCN, and cannot be redefined as identifiers by the test suite writer.

The logical structure of the test suite is represented in TTCN.GR by graphical elements such as boxes, headers, lines of text and indentation, and in TTCN.MP by BNF keywords, by position in the string and by indentation level numbers. We will illustrate the correspondences below.

The syntax of TTCN.GR is described by table patterns and by natural language. Table patterns include BNF rules for table entries. BNF means *Backus-Naur form* or *Backus normal form* and is a family of notations for regular expressions. BNF rules are also called *productions*. The syntax of TTCN.MP uses only BNF rules and natural language. Both the tables and table entries in TTCN.GR, and the productions of TTCN.MP, are restricted by the same set of *static semantic rules*. They deal with contextual constraints, which it would be overly complicated or impossible to specify by means of BNF. We will deal with static semantic rules in the context of the tables to which they apply. However, we will not particularly mention static semantic rules that are common to programming languages, such as the following:

- A type must be defined before data objects of this type are declared.
- A data object must be declared to be of a certain type before it is assigned a value.
- A data object must have a value assigned to it before it is "read."
- Values of data objects must be compatible with the declared types as well as with the operations executed on them.
- In a "function or procedure call" of any kind, the number and types of actual parameters must agree with those of the formal parameters in the declaration.

A complete list of TTCN BNF and static semantic rules is provided in Annex A.

3.2.1 BNF metarules

BNF metarules describe the syntax and semantics of the BNF rules, i.e. the BNF variant used by TTCN. The standard uses separate lines, small print and rule numbers to distinguish BNF rules from ordinary text; we will omit the rule numbers and use italics for alphanumeric characters.

Every possible TTCN test suite can theoretically be created by a production process consisting of the following steps:

- Production starts with the symbol string consisting of the nonterminal symbol *Suite*.
- Every nonterminal symbol, rendered in BNF rules in *lightface italics*, will appear as the left-hand side, i.e. to the left of the symbol "::=", of exactly one BNF rule. An occurrence of the nonterminal in the current string can be replaced by the right-hand side of this rule. Right-hand sides are symbol strings formed from nonterminal and terminal symbols. Terminal symbols appear in TTCN BNF rules in ***boldface italics*** or enclosed by quotation marks ("...").
  Suite, for example, can be replaced according to the rule

  Suite ::= ***$Suite*** *SuiteId SuiteOverviewPart DeclarationsPart ConstraintsPart DynamicPart* ***$End_Suite***

- Replacements continue until no nonterminal is left in the current string.

Static semantics limit the choice of possible replacements. If no more replacements are possible, and static semantic restrictions and other rules of 9646-3 have been respected, the resulting string is a syntactically valid TTCN.MP test suite. Execution of the test suite, however, may reveal dynamic test suite errors. They are discussed in the context of various TTCN tables and language elements.

In order to reduce the size of the BNF rule set, the following grammatical symbols have been used:

( )		grouping symbols
A| *B*| *C*	:	*A* or *B* or *C* (alternatives)
[*A*]	:	0 or 1 instance of *A* (option)
{*A*}	:	0, 1 or several instances of *A* (repetition)
{*A*}+	:	1 or several instances of *A* (repetition)

The use of the vertical bar permits to the use of exactly one rule each for the expansion of every nonterminal in TTCN BNF. With the aid of these constructions, a single rule may expand its left-hand side to a wide variety of symbol strings. For example, the fictitious production

A ::= *B* | ([*C*] {*D*} {*E*}+)

permits replacement of *A* by *B*, *E*, *CE*, *CDDE*, or *DEEE*, etc.

In some places we expand the right-hand sides of TTCN BNF rules partly, saving our readers a few production steps. The resulting rules are marked by an asterisc at the end of the nonterminal being expanded. To use a fictitious example, instead of

Greeting ::= *Hi_Greeting* | *Hello_Greeting*
Hi_Greeting ::= ***Hi*** | ***Hi*** "," ***there***
Hello_Greeting ::= ***Hello*** | ***Hello*** "," ***everybody***

we might write

*Greeting** ::= ***Hi*** ["," ***there***] | ***Hello*** ["," ***everybody***].

Without explicit mention in the standard, successive reserved words and identifiers are obviously separated from one another by blanks. Otherwise, for example, NOTABLE would be ambiguous if both ABLE and NOTABLE were Boolean variables.

3.2.2 Single object tables

The typical form of a TTCN.GR single object table is shown in Table 3.1. It defines one abstract "object," e.g. a data type or a pattern for data objects of that type.

TABLE 3.1. The typical form of a single object table in TTCN.GR

<table>
<tr><td colspan="5">TITLE OF SINGLE OBJECT TABLE</td><td>TITLE</td></tr>
<tr><td colspan="5">OBJECT NAME : Object Identifier
FURTHER INFO : ...
:
COMMENTS : There need not be any text here.</td><td>header</td></tr>
<tr><td>COLUMN 1</td><td>...</td><td>COLUMN N</td><td>...</td><td>COMMENTS</td><td rowspan="2">body</td></tr>
<tr><td>Entry 1.1
:
Entry m.1
:</td><td></td><td>Entry 1.n
:
Entry m.n
:</td><td></td><td>This column may be omitted.</td></tr>
<tr><td colspan="5">DETAILED COMMENTS: This footer may be omitted.</td><td>footer</td></tr>
</table>

General information about this object, in its entirety, can be found in the header, including an optional short general comment. The table body contains a row of (one or more) column titles and (one or more) rows of column entries, the last entry being an optional comment to the given row. It seems that in practice, if there is no comment entry in the header, some test suite writers tend to omit the title "COMMENTS:", too, although the standard does not authorize this. A comments column consisting only of the column title "COMMENTS" may, however, be omitted completely, which then leaves more space for the other columns. The optional footer is provided for more extensive comments, and may be referenced from the shorter comments in the header or the body of the table.

An example of a single object table is the structured type definition, described in Table 3.10, and for which a concrete example is shown in Table 3.11. Other examples are ASN.1 type definitions, which have only one column with only one entry, cf. Table 3.13. Note that this single entry may spread graphically over several lines of text. Line continuation is addressed explicitly in the context of behavior descriptions, where readability problems are aggravated by the significance of indentation, cf. 4.2.6. In other tables, the handling of line continuation is left to common sense and, where applicable, to ASN.1 conventions (which is not meant to imply any conflict between the two).

The single object table schema of Table 3.1 corresponds to the BNF rules schema in Table 3.2. This example illustrates the use of "*Comment*", "***$***", "***Begin_***", and "***End_***" in the TTCN BNF rendition of table structure Other actual names, titles and entry nonterminals will, of course, differ from those shown and will depend on the particular object described in the table. Not only entry rows (cf. *Row* in Table 3.2) and groups of rows, but also specific groups of tables, up to the entire test suite, are always expanded as follows:

SomeName ::= ***$SomeName*** ... ***$End_SomeName***

TABLE 3.2. BNF rule schema for a single object table

Title_of_Single_Object_Table ::= ***$Begin_Title_of_Single_Object_Table*** *ObjectName FurtherInfo* ... [*Comment*] *Rows* [*Comment*] ***$End_Title_of_Single_Object_Table***
ObjectName ::= ***$ObjectName*** *ObjectIdentifier*
FurtherInfo ::= ***$FurtherInfo*** *FurtherInfoContents*
... etc. ...
Comment ::= ***$Comment*** [*BoundedFreeText*]
Rows ::= ***$Rows*** {*Row*}+ ***$End_Rows***
Row ::= ***$Row*** *Column_1_Entry* ... *Column_n_Entry* ... [*Comment*] ***$End_Row***
Column_1_Entry ::= ***$Column_1_Entry*** *C_1_E_Contents*
... etc. ...

3.2.3 Multiple object tables

The typical form of a TTCN.GR multiple object table is shown in Table 3.3. A multiple object table describes a set of objects with identically structured information, for example the set of test case variables. In this case, the table body contains one row of entries for each object. Each row consists of a list of entries, the leftmost being the object's name, the rightmost being optional comments for the object. The comments column and the footer are optional, as in single object tables.

TABLE 3.3. The typical form of a multiple object table in TTCN.GR

TITLE OF MULTIPLE OBJECT TABLE					
OBJECT NAMES	...	COLUMN N	...	COMMENTS	body
Object Identifier 1 : Object Identifier m :		: : : :		This column can be omitted.	
DETAILED COMMENTS : This footer can be omitted.					footer

(title)

An example of a multiple object table is the list of test suite variable declarations, defined in Table 3.19, for which a concrete example is shown in Table 3.20.

The multiple object table schema of Table 3.3 corresponds to the BNF rules schema in Table 3.4. Again, actual names, titles and entries will depend on the particular objects described in the table.

TABLE 3.4. BNF rule schema for a multiple object table

Title_of_Multiple-Object_Table ::= ***$Begin_Title_of_Multiple_Object_Table***
    {*ObjectInfo*}+ [*Comment*] ***$End_Title_of_Multiple_Object_Table***

ObjectInfo ::= ***$ObjectInfo*** *ObjectName* ... *Column_n_Info* ... [*Comment*]
    $EndObjectInfo

ObjectName ::= ***$ObjectName*** *ObjectIdentifier*

... etc. ...

Column_n_Info ::= ... etc. ...

... etc. ...

Comment ::= ***$Comment*** [*BoundedFreeText*]

Some tables do not follow either of these two schemes. Specifically, there exist ways in TTCN.GR to merge several single object tables into one compact table. In these cases, the corresponding TTCN.MP is based on the original set of tables. No merged form has been defined for TTCN.MP.

3.2.4 Some important syntactical elements

In table patterns and BNF rules some basic nonterminals appear repeatedly. We will describe these now.

Many object names are *identifiers*. Identifiers start with an upper or lower case letter, followed by any number and combination of letters, digits and underscores:

*Identifier** ::= *Alpha* {*Alpha* | *Num* | "_"}
*Alpha** ::= ***A*** | ***B*** | ... | ***Z*** | ***a*** | ***b*** | ... | *z*
*Num** ::= ***0*** | ***1*** | ***2*** | ***3*** | ***4*** | ***5*** | ***6*** | ***7*** | ***8*** | ***9***

Many specific identifiers in TTCN tables, such as *SuiteIdentifier, TestGroup-Identifier,* etc. are BNF nonterminals ending in "*Identifier*". Each of these expands by a BNF rule to *Identifier,* for example

SuiteIdentifier ::= *Identifier*

Informal elements of a test suite, in particular comments, but also refences to standards, test method descriptions, test purposes, informal operation definitions, etc., appear as *bounded free text* or *free text*. Bounded free text is free text enclosed by "comment delimiters":

BoundedFreeText ::= "/*" *FreeText* "*/"

Free text consists of a character string composed of characters specified in the ISO 10646 alphabet. As it may contain TTCN.MP keywords and other delimiters, it would generally not be possible to determine the end of free text in TTCN.MP. Therefore, entries consisting of free text in TTCN.GR are represented as bounded free text in TTCN.MP. In the free text string, all occurrences of "*/" in free text are replaced by "*/", and all occurrences of "\" in free text are replaced by "\\", both in one pass. This way, any string can be made into TTCN.MP bounded free text without danger of readers mistaking an occurrence of "*/" within the string for the closing delimiter at the end of the bounded free text. However, in graphical TTCN, table layout provides already for a clear delimitation of free text. Hence, wherever bounded free text occurs in TTCN.-MP, the corresponding entry in the respective table in TTCN.GR contains only the original free text, without comment delimiters and without any recoding of "*/" or "\".

The arguments of any type of functional expression are regularly given in the form of a typed formal parameter list, or, in short, a *parameter list*. Its syntax is determined by the following BNF rules:

> *FormalParList* ::= "(" *FormalPar&Type* {";" *FormalPar&Type*} ")"
> *FormalPar&Type* := *FormalParIdentifier* {"," *FormalParIdentifier*} ":" *FormalParType*
> *FormalParType* ::= *Type* | *PCO_TypeIdentifier* | ***PDU***

Using this notation, one can write single arguments, e.g. (P1:TypeA), single type argument lists, e.g. (P1,P2,P3:TypeA), and multi-type argument lists, e.g. (P1:TypeA; P2,P3:TypeB). One can even write syntactically different but semantically equivalent formal parameter lists, e.g. (A:INTEGER; B:INTEGER; C:BOOLEAN) and (A,B:INTEGER; C:BOOLEAN).

Expressions are another element occurring in many places. An expression is composed of at least one basic expression (BNF nonterminal *Primary*) and zero or more TTCN operators and pairs of parentheses. Basic expressions are either explicit values (*Value*), data object names or components of structured data objects (*DataObjectReference*), functional expressions (*OpCall*), selection expression names (*SelectExprIdentifier*), or expressions enclosed in parentheses. Parentheses are used to enforce the desired order of evaluation of operators. If *ex1* is an expression and *unop* is a unary operator applicable to *ex1*, then *unop ex1* is an expression. If *ex1* and *ex2* are expressions and *binop* is a binary operator applicable to *ex1* and *ex2*, then *ex1 binop ex2* is an expression. Applicability refers to issues of type, for example, that expressions with Boolean values are not added like integers.

Expressions usually evaluate to specific values of given types. Subsection 3.5.1 deals with possible exceptions, namely the case that constraint values are used as expressions.

3.2.5 Redundancy of the modes of expression in TTCN

Attentive readers will discover on many occasions that a test case can be written in many different, but basically equivalent, ways, without changing its verdict assignment or its behavior with the IUT. Test suite writers often have a choice between a number of equally appropriate means of expression. They can choose according to their liking, advisably, of course, in a consistent style. This concession to the different personal tastes of test suite writers complicates both TTCN and, more importantly, the implementation of TTCN tools. This redundancy can be attributed to the fact that during the early stages of TTCN development, standardization of sizeable test suites was attempted, based either on preliminary versions of TTCN, or on a concept of what TTCN would be like. This led to diverging styles, many of which were taken into account in the progression of the standard. Readers may know of other instances of this compromise effect on standards.

3.3 TTCN test suites

3.3.1 The parts of a TTCN test suite

A TTCN test suite consists of four parts:

- the *Test suite overview part*, which
  - identifies relevant protocol or profile standards,
  - includes a commented overview of the hierarchical group structure of the test suite, and
  - contains commented indices of the test cases, test steps, and default behaviors,
- the *declarations part*, in which the data types, data objects and other components relevant to the test suite are defined and declared, respectively,
- the *constraints part*, in which the concrete data objects and patterns for data objects associated with send and receive events are defined, as well as
- the *dynamic part*, in which the
  - test cases

  associated with the standardized catalog of test purposes and the chosen test method are described. This part also contains a

  - test step library and a
  - default behavior library,

  in which parts of the behavior descriptions, common to several test cases, are collected. The name "dynamic part" does not imply that this part is to be considered as variable. Rather, it contains the (fixed) descriptions of the dynamics of the test cases. TTCN users often call it informally the *behavior part*.

This order of the test suite parts is fixed. As a supplement to the declarations part, a test suite may contain an ASN.1 module as an appendix.

3.3.2 The test suite overview part

The test suite overview consists of the test suite structure, test case index, and, optionally, test step index and default index tables.

The *test suite structure table* conveys in its header general information concerning the test suite, as shown in Table 3.5.

TABLE 3.5. Test suite structure

TEST SUITE STRUCTURE			
SUITE NAME : *Identifier* – the name of the test suite STANDARDS REF : *FreeText* – references the relevant protocol and service standards PICS REF : *FreeText* – references the PICS proforma PIXIT REF : *FreeText* – references the PIXIT proforma TEST METHOD(S) : *FreeText* – identifies the abstract test method(s) for which the test suite is intended; includes reference to a test management protocol, if the ATM is "coordinated" COMMENTS : [*FreeText*]			
TEST GROUP REFERENCE	SELECTION REF	TEST GROUP OBJECTIVE	PAGE NR
:	:	:	:
TestGroupReference	[*SelectExprIdentifier*]	*FreeText*	*Number*
:	:	:	:
DETAILED COMMENTS: [*FreeText*]			

While there is only a limited choice between four standardized abstract test methods in the single-party context (cf. Section 2.5), TTCN is not very explicit about how to specify test methods in the multi-party context. A reasonable choice might be to use the names of all test component configurations, if there are not too many.

The body of the test suite structure table consists of an annotated index of test groups in their "natural order", i.e. depth first in their hierarchy tree. A test group reference is a path name, composed of all supergroup names in hierarchical order, each followed by a slash symbol:

TestGroupReference ::= [*SuiteIdentifier* "/"] {*TestGroupIdentifier* "/"}

The optional selection expression identifier is used if it controls the selection of all test cases in this group for execution, that is to say, if the expression describes a necessary precondition for the execution of these test cases. The selection expression identifier must reference a Boolean expression defined in the test case selection expression definitions table, cf. 3.4.5. Test group objectives have been addressed in 2.6.3. Page number entries refer to the page on which the first test case of the given group is printed in the ATS. They are not included in TTCN.MP.

Note our representation of test suite table patterns:

- Standardized titles are written in MIXED CASE CAPITALS.
- Nonterminals for table entries, the syntax of which is defined in BNF, are written in *italics*.
- Semantic comments to entries are written in normal font.
- BNF metasyntax is occasionally used within table patterns; e.g. optional entries are indicated by [*square brackets*].

There are no requirements in [ISO 9646] as to the actual font(s) to be used in an ATS.

The *test case index*, the pattern of which is presented in Table 3.6, is a multiple object table listing all test cases in the order in which they appear in the dynamic part. The test cases are ordered by test groups, which themselves are arranged as in the test suite structure table.

TABLE 3.6. Test case index

TEST CASE INDEX				
TEST GROUP REFERENCE	TEST CASE ID	SELECTION REF	DESCRIPTION	PAGE NR
:	:	:	:	:
[*TestGroupReference*]	*TestCaseIdentifier*	[*SelectExprIdentifier*]	*FreeText*	*Number*
the path name of the lowest level test group containing the test case; given only with the first TC of the group	the name of the test case	see 3.4.5	informal , e.g. a short description of the test purpose	
:	:	:	:	:
DETAILED COMMENTS: [*FreeText*]				

Test steps are common parts of test cases, collected in a specific library in the dynamic part, cf. 2.6.3, 4.1.9 and 4.2.4. They have much in common with test cases, and their index strongly resembles the test case index. Its pattern is presented in Table 3.7. Whether a test step is executed depends solely on the execution of the test cases able to call it. Hence, unlike test cases, test steps are neither selected nor deselected, nor are they associated with selection expressions.

TABLE 3.7. Test step index

TEST STEP INDEX			
TEST STEP GROUP REFERENCE	TEST STEP ID	DESCRIPTION	PAGE NR
: [*TestStepGroupReference*] :	: *TestStepIdentifier* :	: *FreeText* :	: *Number* :
DETAILED COMMENTS: [*FreeText*]			

Let us abbreviate "default (dynamic) behavior" informally by "default." Defaults are also common parts of test cases, collected in a specific library in the dynamic part, cf. 4.5. Their index is virtually identical in form with the test step index, as shown in Table 3.8.

TABLE 3.8. Default index

DEFAULT INDEX			
DEFAULT GROUP REFERENCE	DEFAULT ID	DESCRIPTION	PAGE NR
: [*DefaultGroupReference*] :	: *DefaultIdentifier* :	: *FreeText* :	: *Number* :
DETAILED COMMENTS: [*FreeText*]			

3.4 Definitions and declarations

Before going into the TTCN specific details, it might be in order to spend a few words on terminology we use in connection with data. Space considerations and the practical focus of this book forbid a formal treatment of data types, but we think a few informal examples may already help to clarify the discussion.

In TTCN, we are dealing with concrete data types, i.e. types which are constructed from elementary sets, called simple types, like integers and bitstrings. The constructions determine which operations are applicable to data objects of the constructed types, but we will not go into any appreciable details there.

A concrete *data type* is similar to a set of *data objects*, also called *values*. We do not say "identical," because a TTCN data type definition will, for example, not only specify indirectly a Cartesian product of sets, but also what the components of its elements are called. In fact, accessing a component by its name is an example of an operation applicable to data objects of this type.

In a type definition, we can define that a data object of *ExampleType* has a first *component*, called *comp1*, of type INTEGER, and a second component, called *comp2*, of type PrintableString. Then "*Obj*:=(1,"Testing,Testing")" is an intuitive *specification* for the unique data object *Obj* of *ExampleType* having the property that *comp1* is 1, and *comp2* is "Testing, Testing". Note that we have not used actual TTCN syntax.

In a similar way as in pattern matching languages, we may define specific *patterns* for data objects of *ExampleType* by requiring that, for example, in order to match *ExamplePat*, *comp1* must be different from 5, and *comp2* must consist of up to 20 characters. Our example data object *Obj* matches *ExamplePat.* Identity to *Obj* is another pattern. It is matched only by *Obj*.

Again, we have not used TTCN syntax yet. TTCN has very detailed syntactic and semantic rules on how to specify data types, data objects and patterns, and it will take a few sections to present these.

We will use the term *structure* in an informal sense. Data objects having components have a structure. Their structure is by definition more complex than the structure of their components. They can be represented as trees.

3.4.1 The description of data objects and patterns for data objects

Each test case in the dynamic part of a TTCN test suite consists primarily of a description of possible sequences of events at interaction points

- between test components, i.e. at CPs,
- between the tester on one side and the IUT or the underlying provider on the other side, i.e. at PCOs, and
- between the tester and its timers.

For the sake of readability, the data objects sent and received in such events at CPs and PCOs, e.g. ASPs, PDUs and CMs, as well as patterns for these data objects, are taken out of the behavior descriptions and collected separately. They are defined and named in the constraints part, as are their relevant components. They are then referenced by name from the dynamic part.

Prior to the specification of data objects and patterns in the constraints part, the associated data types are defined in the *declarations part.* This latter test suite part is also used for other declarations; in all, it contains definitions and declarations of

- test suite specific general data types and data operations,

- test suite parameters, i.e. protocol parameters which obtain their values from ICS and IXIT entries,
- test case selection expressions, used to determine which test cases are applicable based on ICS and IXIT entries,
- constants and variables, more precisely, test suite constants, test suite variables and test case variables,
- PCOs, the points of interaction (control and observation) between the tester and the IUT or the underlying service provider,
- CPs, the points of interaction between different test components,
- timers,
- test components and their configurations,
- ASP, PDU and CM types, i.e. the data objects exchanged at PCOs and CPs, possibly including their encoding, as well as of
- aliases, i.e. substitute names that can be used as abbreviations in test case behavior descriptions.

The test relevant patterns of ASPs, PDUs, CMs and their components, called *constraints*, are defined in the constraints part. These data patterns are defined by filling the components of the respective data types with concrete values or with value patterns, expressed in a rich assortment of data matching mechanisms. Note that there is no strict boundary between data types, data patterns and data objects in TTCN. A data pattern specifies the set of matching data objects from a previously defined data type or data pattern. In a way, it defines a subset, similarly to the way some type definitions do. In extreme cases, there is exactly one data object which matches the given data pattern or data type. The overlap between type and constraint mechanisms is felt particularly strongly in observing that narrowing the range of the integers can be achieved in

- a type definition by restriction of a predefined type, see Table 3.9 and the end of Subsection 3.4.2,
- a type definition by restriction of a TTCN-defined type, see 3.4.2,
- a constraint definition with a value range, see Section 3.5,
- a modified constraint, see Section 3.5, or in
- any combination of the above.

Again, it is the test suite writer's responsibility to choose a clear style and to stick with it.

3.4.2 Data types

Some data types are predefined in TTCN, while others are defined specifically for a test suite by the restriction of value ranges and by the formation of structures. Moreover, ASN.1 is fully integrated into TTCN: ASN.1 type definitions can be specified explicitly within the test suite, in particular ASN.1 oriented TTCN tables, and they can be imported, e.g. from a protocol standard. There exist very readable tutorial texts on ASN.1. The standard itself [ISO 8824] contains tutorial passages. A thorough presentation is given in [STEE 990]. Shorter introductions include the relevant chapters in [ROSE 90] and, as a very concise survey, [RUSS 89], to name but a few.

PREDEFINED DATA TYPES

TTCN adopted the following simple data types from ASN.1 [ISO 8824]:

- INTEGER, with allowed values of -17,0 , 256, etc.
- BOOLEAN, with values TRUE or FALSE
- BITSTRING, single-quoted strings of 0's and 1's, followed by the character "B", such as '010011'B
- HEXSTRING, single-quoted strings of the figures 0 – 9 and the letters A – F, followed by the character "H", e.g. '02BAD4DAD'H
- OCTETSTRING, single-quoted strings of even length of the figures 0 – 9 and the letters A – F, followed by the character "O", e.g. 'AFABCAFE4ABE'O
- the character string types NumericString, PrintableString, TeletexString, VideotexString, VisibleString, IA5String, GraphicString, GeneralString – enclosed in string quotes " " ", e.g. "TTCN". A string quote occurring as a character within a string is written as " "" ", i.e. a pair of string quotes.

The value notation that TTCN defines for these data types agrees with the corresponding notation in ASN.1, cf. the examples given above as well as the following: *Number* is the notation for non-negative integers, used for example by ASN.1 in the definition of integer notation and by TTCN in the definition of integer range boundaries. The ASN.1 standard says "A number shall consist of one or more digits. The first digit shall not be zero, unless the number is a single digit." TTCN gives the following equivalent BNF rules:

Number ::= (*NonZeroNum* {*Num*}) | ***0***

NonZeroNum ::= ***1*** | ***2*** | ***3*** | ***4*** | ***5*** | ***6*** | ***7*** | ***8*** | ***9***

Num ::= ***0*** | *NonZeroNum*

By the embedding of ASN.1 into TTCN, other data types, like REAL, are available. In the context of protocols they are not needed very often.

The second version of TTCN will include the predefined *R_Type* with the possible values none, pass, inconc, and fail. These values are used to store and transfer verdicts, for example, in variables and messages.

THE TABULAR DEFINITION OF ADDITIONAL DATA TYPES

Data handling in TTCN is possible both without the use of ASN.1, in a TTCN specific purely tabular manner, and in tables using ASN.1. We start with the purely tabular variant.

Test suite specific data types can be created from previously defined ones (predefined or user defined) either by restriction or by the formation of structures. Type definitions by the restriction of the integer value range or of the string length are given all together in one table, which is of the form shown in Table 3.9.

TABLE 3.9. Type definitions by restriction

SIMPLE TYPE DEFINITIONS		
TYPE NAME	TYPE DEFINITION	COMMENTS
: *SimpleTypeIdentifier* the name of the new data type :	: *Type&Restriction* the name of the data type upon which the new type is based, plus an optional (range restriction) or a (value list) or a [length restriction] :	: [*FreeText*] :
DETAILED COMMENTS: [*FreeText*]		

Of course, the restrictions must be applicable to the types to be restricted, as in the following examples of *Type&Restriction*, where the type definitions are shown on the left-hand side and the corresponding value sets are described on the right-hand side:

INTEGER	{ ..., -1, 0, 1, 2, ... }
INTEGER(0,1,2,3,4)	{0, 1, 2, 3, 4}
INTEGER(0 .. 1023)	{0, 1, 2, ... ,1023}
INTEGER(0,INFINITY)	{0, 1, 2, ... }
IA5STRING[10]	IA5 strings, exactly of length 10
HEXSTRING[4 .. 8]	Hexstrings having a length from 4 through 8

The first example shows that it is not mandatory to give a restriction, i.e. type definitions can be used for renaming. Integer and length restrictions can be used in other tables, too, and are discussed in more detail later in this subsection.

Structured data types can also be defined in tabular form. Being constructed from a fixed sequence of named components of various types, they resemble what are called lists, tables or records in some programming languages, or SEQUENCEs in ASN.1. The table pattern for structured type definitions is shown in Table 3.10.

TABLE 3.10. Structured type definition

STRUCTURED TYPE DEFINITION		
TYPE NAME : *StructId&FullId* – the type name, if necessary followed by the standardized type name in parentheses COMMENTS : [*FreeText*]		
ELEMENT NAME	TYPE DEFINITION	COMMENTS
: *ElemId&FullId* the element name, if necessary followed by the standardized name in parentheses :	: *Type&Attributes* the element type, optionally followed by a length restriction. Possible entry: "PDU". :	: [*FreeText*] :
DETAILED COMMENTS: [*FreeText*]		

The type of an element may itself be a structured type, such that tree structures of arbitrary finite depth can be defined. Contrary to ASN.1 type definitions, recursion is not allowed, i.e. a simple or structured type definition must not refer to itself in the type definition column, not even indirectly via other type definitions. Type definitions like "A linear list of integers is either *Empty,* or it has two components: *c1*, of type INTEGER, and *c2*, which is a linear list of integers," cannot be expressed by means of structured type definitions.

All elements of a structured type are automatically considered as optional. This convention may force test suite writers to check for the presence of mandatory elements explicitly, e.g. by using the operation IS_PRESENT, or by defining appropriate constraints. The pseudotype "PDU" in a type element represents the union of all PDU types declared for the test suite. In a constraint of the structured type in question, this element is replaced by some PDU type constraint. Constraints are covered in Section 3.5.

Let us consider in Table 3.11 a simple example of a structured type definition, which we will later contrast with a corresponding ASN.1 type definition.

TABLE 3.11. A structured type definition example

STRUCTURED TYPE DEFINITION		
TYPE NAME : BodyHeight		
ELEMENT NAME	TYPE DEFINITION	COMMENTS
feet inches	INTEGER_1_THRU_8 INTEGER_0_THRU_11	assumed to be defined as a simple type assumed to be defined as a simple type

THE ASN.1 DEFINITION OF ADDITIONAL DATA TYPES

ASN.1 is embedded in TTCN; it can be used in test suite specific general type definitions. In that case each type is defined in a separate table of the form shown in Table 3.12. Only ASN.1 external type references must not be used.

TABLE 3.12. ASN.1 type definition

ASN.1 TYPE DEFINITION
TYPE NAME : *ASN1_TypeId&FullId* – the type name, followed by the standardized type name in parentheses, if necessary COMMENTS : [*FreeText*]
TYPE DEFINITION
ASN1_Type&LocalTypes – consisting of one *Type* and any number of *TypeAssignments* as supported by ASN.1. Moreover, comments commencing with (or enclosed by) "--" may be used as in ASN.1.
DETAILED COMMENTS: [*FreeText*]

There are some advantages to the use of ASN.1. First of all, many type definitions can be taken directly from the protocol standards. Furthermore, ASN.1 permits the definition of types of unbounded and variable size (SET OF, SEQUENCE OF, recursion), while TTCN tabular type definitions describe only types of a fixed tree structure with optional branches, cf. the remarks about Table 3.10.

The type assignments following the principal *Type* in the "Type Definition" field play the role of local auxiliary types. They may refer to each other within this definition, but they cannot be referenced from outside of this table. Types referred to from the type definition may be either local auxiliary types, defined in other ASN.1 type definitions, or imported by an ASN.1 type definition by reference (see below).

In identifiers occurring in TTCN tables using ASN.1, the dash (hyphen) character "-" must not be used. The underscore "_" may be used instead. As typographical line breaks do not show up in TTCN.MP, ASN.1 comments ending with the end of the line instead of a double hyphen, though harmless in TTCN.GR, may cause parsing problems in TTCN.MP.

Let us contrast the tabular definition of the BodyHeight type with the ASN.1 definition of the type *Size* given in Table 3.13. In the latter we have illustrated various ways of procuring ASN.1 subtypes: from different ASN.1 type definitions (explicit or by reference) and from among local auxiliary types.

TABLE 3.13. An ASN.1 type definition example

ASN.1 TYPE DEFINITION
TYPE NAME : Size
TYPE DEFINITION

SEQUENCE {
ft FEET_Type ,
in INCHES_Type
}
FEET_Type ::= INTEGER { minimum(1), maximum(8) }
-- INCHES_Type is assumed to be defined in another ASN.1 type definition. --

ASN.1 data types may be imported from outside of the ATS, usually from a service or protocol specification, or from an ASN.1 module included with the ATS. All these imports are specified in one table of the form shown in Table 3.14. In the course of this import, all dash characters "-" in ASN.1 identifiers are changed to underscores "_", such that all identifiers become TTCN identifiers.

INTEGER RANGES AND LENGTH RESTRICTIONS

Data types describing the entire set or a bounded interval of integers can be used to define new data types by restricting the values to an integer range. Constraint entries for an integer value also can be defined as an integer range. Integer ranges appear as *IntegerRange* in the expansion of the BNF nonterminal *Type&Restriction* and as *ValueRange* in the expansion of the BNF nonterminal *ConstraintValue.* In both cases they have the form

(min .. max) or (min TO max).

TABLE 3.14. ASN.1 type definition by reference

ASN.1 TYPE DEFINITIONS BY REFERENCE			
TYPE NAME	TYPE REFERENCE	MODULE IDENTIFIER	COMMENTS
:	:	:	:
ASN1_Type&FullId	*TypeReference* a "typereference" in ASN.1	*ModuleIdentifier* as defined in ASN.1	[*FreeText*]
:	:	:	:
DETAILED COMMENTS: [*FreeText*]			

Both bounds must be, or evaluate to, integers such that min < max. The lower and upper bound may also be -INFINITY and INFINITY, respectively, denoting the absence of a bound in this direction. In simple type definitions, the restricting range must be a subrange of the restricted type's range. In a constraint, a range entry may be chosen arbitrarily, though it cannot be recommended that this capability be used to increase an integer range.

Data types or constraints consisting of a number of uniform elements can be used to define new data types by restricting the number of elements in data objects, often called their *length*. Length restrictions may be applied to basic string types, whose characters are considered to be the elements, or to ASN.1 SET OF or SEQUENCE OF types. They cannot be used with integers.

Length restrictions appear in the expansion of the BNF nonterminal *Type&Restriction* as *LengthRestriction,* within *Type&Attributes* as *LengthAttribute,* and within *ConstraintValue&Attributes* as *ValueLength.* Length restrictions may have the following forms:

- [length], which gives a required exact length, and
- [min_length TO max_length] or [min_length .. max_length],

  which defines a range of permitted lengths. Length numbers (minimal, maximal or exact) must be or evaluate to nonnegative integers such that min_length < max_length. The latter may also be "INFINITY", denoting the absence of an upper bound.

Probably unintentionally, TTCN permits even negative numbers and - INFINITY in *LengthRestrictions*. It might be advisable to avoid these values, which even permit type definitions that specify the empty set of data objects, such as BITSTRING[-2,-1].

Length restrictions must always specify sets of data objects which are subsets of the type being restricted, both in type definitions and in constraint declarations. This rule does not exclude the empty set.

3.4.3 Operators and operations

In TTCN, the following distinction is made between operators and operations: Operations have names that are TTCN identifiers (cf. 3.1.4), and are written in the usual function notation "operation_name(parameter_list)." On the other hand, operators may contain special symbols and may be used in special positions, as, for example, one usually writes "a+b" instead of "Plus(a,b)". TTCN operators and basic TTCN operations are predefined in TTCN. Additional test suite specific operations may be defined in the declarations part.

OPERATORS

The customary *arithmetic*, *relational* and *Boolean operators* are available in TTCN.

The arithmetic operators

+ - * / MOD

are used on integers, i.e. on expressions of types that have been defined by restriction(s) of INTEGER. Both - and (even) + may also be used in unary position.

Out of regard for possible limitations in available character sets, ≤, ≥ and ≠ are written as <=, >= and <>, respectively. The relational operators

= < > <> <= >=

can all be applied to integers. The operators <> and = can even be applied to non-numeric operands, but the types must be compatible - a restriction which is not clearly defined.

The Boolean operators

NOT AND OR

are used as in propositional logic. Both arguments of AND or OR must be defined. For example, the Boolean term "FALSE AND some_proposition," with some_proposition undefined, is really undefined itself and does not evaluate prematurely to FALSE; its evaluation leads to a test case error.

As an example containing all three kinds of operators, the Boolean expression

('FF'O <> '255'O) AND ((-0) >= (129 MOD 128))

will evaluate to FALSE.

As usual, precedence rules save parentheses. The precedence rules of TTCN operators are shown in Figure 3.3. Though similar to those of some programming languages, they are not quite optimal. It might have been more useful not only to give arithmetic operators precedence over relational operators, but also to give both arithmetic operators and relational operators precedence over Boolean operators, because this corresponds to the way expressions are usually nested.

general grouping	(...)				
Boolean		NOT	AND	OR	
arithmetic		+... −...	* / MOD	...+... ...−...	
relational					= < > <> <= >=
	high ◄——		priority	——►	low

FIGURE 3.3. Precedence among TTCN operators

Chained binary operators of equal priority evaluate from left to right. Readability will sometimes improve if the dropping of parentheses is not carried too far; for example compare the following two expressions with the same value:

9 - ((3 * (- 10)) MOD 14) + 3

9 - 3 * - 10 MOD 14 + 3

PREDEFINED OPERATIONS

TTCN provides predefined conversion operations and data object analysis functions. Conversion operations convert between INTEGER values on the one hand and HEXSTRING or BITSTRING values on the other. The operation names indicate the direction of the conversion.

HEX_TO_INT converts a HEXSTRING expression to an INTEGER value, namely, to the value obtained if the argument is interpreted as a hexadecimal number, e.g.

HEX_TO_INT('001FF'H) = 511.

INT_TO_HEX converts, roughly speaking, an INTEGER expression to a HEXSTRING value. More precisely, it maps a pair (*integer,length*) of two IN-

TEGER expressions of nonnegative value to a HEXSTRING value, namely the hexadecimal number with the value of *integer*, padded to the left with zeros, such that altogether *length* characters are obtained. The value of *length* must be sufficiently large. Examples:

INT_TO_HEX(511,7) = '00001FF'H
INT_TO_HEX(511,2) is invalid.

For both operations, ''H is a borderline case: it is not clear from the standard whether HEX_TO_INT(''H) or INT_TO_HEX(0,0) is defined.

BIT_TO_INT converts a BITSTRING expression to an INTEGER value, namely, to the value obtained if the argument is interpreted as a binary number, e.g.

BIT_TO_INT('000101'B) = 5.

INT_TO_BIT converts, roughly speaking, an INTEGER expression to a BITSTRING value. More precisely, it maps a pair (*integer*,*length*) of two INTEGER expressions of nonnegative value to a BITSTRING value, namely the binary number with the value of *integer*, padded to the left with zeros, such that altogether *length* characters are obtained. The value of *length* must be sufficiently large. Examples:

INT_TO_BIT(5,6) = '000101'B
INT_TO_BIT(5,2) is invalid.

Similarly as above, it is not clear from the standard whether BIT_TO_INT(''B) or INT_TO_BIT(0,0) is defined.

If an integer argument is negative or the length provided is insufficient, the application of a conversion operation results in a test case error. This is a special case of the general rule that the attempt to do something in the execution of a test suite that is explicitly not permitted leads to a *test case error*. As in programming languages, one cannot preclude by previous static analysis that errors of this kind will occur. Test case errors are recorded in the test log and lead to the immediate termination of the current test case. Since the resumption of testing with a new test case calls for a stable initial state in the tester and in the IUT, operator intervention will usually be required.

The other four predefined operations serve in the analysis of data objects. Let "object_name.element_name" be the identifier of an optional or ASN.1 DEFAULT element (or a field or a parameter) of a data object, written in dot notation. The presence check

IS_PRESENT(object_name.element_name)

indicates whether the element is actually present in the object by returning TRUE or FALSE. If, for example, SomeHeight = (6,–) is an object of type BodyHeight, cf. Table 3.11, then IS_PRESENT(SomeHeight.inches)=FALSE.

The element count operation NUMBER_OF_ELEMENTS, applied to a data object declared as ASN.1 SET OF or SEQUENCE OF, returns the number of elements of that object.

The length count operation LENGTH_OF, applied to a data object of type BIT-, HEX-, OCTETSTRING or some characterstring, returns the number of bits or hexadecimal characters or octets or characters, respectively, in that object. Thus, bearing in mind that one octet is represented by two hexadecimal characters,

LENGTH_OF('11FF'O)=2.

Let object_name be a data object of the type ASN.1 CHOICE, where "identifier_1" precedes one of the types of the choice. Then the choice test operation IS_CHOSEN, applied to the argument "object_name.identifier_1," indicates whether the data object is actually of the type associated with the identifier by returning TRUE or FALSE. If for example SomePDU is of the ASN.1 type

```
CHOICE  {  p1  PDU_Type1
           p2  PDU_Type2
        },
```

then, if SomePDU is actually of type PDU_Type2,

IS_CHOSEN(SomePDU.p2)=TRUE.

This notation and its explanation not only look complex, they even seem to lead to genuine problems: If one identifier appears as a type identifier in several CHOICE type definitions, then the value of an IS_CHOSEN expression may be ambiguous. So what TTCN needs is actually either some kind of ASN.1 type identifier uniqueness requirement, or the name of the CHOICE type as an additional argument, or the choice component type name (PDU_Type2) as an argument, instead of the type identifier (p2). In the last case, the IS_CHOSEN operation would amount to an operation

"IS_OF_TYPE(object_name,type_name),"

which is at present not provided by TTCN.

Figure 3.4 gives an overview of the predefined TTCN operations.

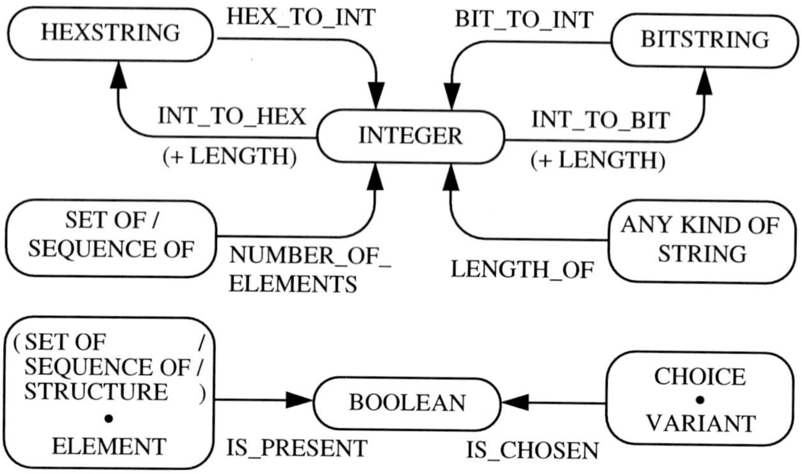

FIGURE 3.4. TTCN predefined operations

TEST SUITE SPECIFIC OPERATIONS

Test suite writers may define additional operations, called *test suite operations*. In the first version of the standard these definitions were given in free text only. Amendment 2 [ISO 9646-3-2] recommends a procedural definition, i.e. a definition in the style of a programming language and with a formal syntax, but it will maintain the possibility of using free text. In the latter case the definition must contain an example of an operation call and the resulting value. The form of a test suite operation definition is shown in Table 3.15.

The usual rule about the agreement of number and values of actual parameters with the number and types of formal parameters applies to calls of operations.

TABLE 3.15. Test suite operation definition

TEST SUITE OPERATION DEFINITION
OPERATION NAME : *TS_OpIdentifier* [*FormalParList*] RESULT TYPE : *Type* COMMENTS : [*FreeText*]
DEFINITION
[*VarBlock*] *ProcStatement* or *FreeText* i.e. either a formal or an informal definition of the operation
DETAILED COMMENTS: [*FreeText*]

Moreover, actual parameters must be bound to values. The effect of an operation call is the provision of a value of the appropriate type in the position of the operation call. Values of actual parameters or other non-local data objects are not changed; actual parameters are passed by value. Side effects will not occur, with one notable exception: Some operation specific memory may be maintained in local variables declared as STATIC in the procedural version. On the first call of the operation, instances of these "persistent" local variables are created within the present test component (if any) or test case. On later calls, their last value in the last operation call is used instead of the declared initial value. Thus, test suite operations using STATIC can be regarded as procedures, objects, or processes, rather than as functions. Of course, this entails the danger of undesired hidden effects, and duplicates memory keeping functions which can be achieved by more explicit means, namely test suite variables.On the plus side of procedural definitions, the obligation to initialize local variables may spare some of the surprises well known from the use of programming languages which do not enforce initialization.

Local variables are declared according to the following syntax:

VarBlock ::= ***VAR*** [*VarDcls*] ***ENDVAR***
*VarDcls** ::= *VarDcl* {";" *VarDcl*}
*VarDcl** ::= [***STATIC***] *VarIdentifier* {"," *VarIdentifier*}":" *Type* ":" *Value*

This syntax is related to that of the formal parameter list, with the STATIC option and the initial values added. The syntax and semantics of the procedure statement follows the usual lines known from programming languages:

ProcStatement ::= *ReturnValueStatement* | *Assignment* |
IfStatement | *WhileLoop* | *CaseStatement* | *ProcBlock*
ReturnValueStatement ::= ***RETURNVALUE*** *Expression*
Assignment ::= *DataObjectReference* ":=" *Expression*
IfStatement ::= ***IF*** *Expression* ***THEN*** *ProcStatement* [***ELSE***
ProcStatement] ***ENDIF***
WhileLoop ::= ***WHILE*** *Expression* ***DO*** *ProcStatement* ***ENDWHILE***
*CaseStatement** ::= ***CASE*** *Expression* ***OF*** *CaseClause* {";"*CaseClause*}
[***ELSE*** *ProcStatement*] ***ENDCASE***
*CaseClause** ::= *IntegerLabel* ":" *ProcStatement*
*ProcBlock** ::= ***BEGIN*** *ProcStatement* {";"*ProcStatement*} ***END***

The return value statement terminates the evaluation of the test suite operation and returns the present value of the expression as the result of the operation. The data object reference in an assignment, i.e. its left-hand side, must denote a local variable or an element of it. This guarantees that external data objects are not affected. The ENDIF was preferred to the block solution of the problem of chained ifs. Case statements must be complete with respect to all possible expression values: The ELSE-part of the case statement is executed if none of the integer labels of the case clauses has the same value as the expression. If, in the latter case, there is no ELSE-part, it is a test case error.

3.4.4 Test suite parameters

Real protocol standards usually involve a set of parameters, such as a maximum window size for unacknowledged messages or a maximum message length. Any implementation of the protocol assigns specific values to these parameters. Additionally, real protocols involve optional and conditional functions, such as error recovery in the Transport protocol. Implementors can choose, either freely or subject to specific conditions, whether to include these protocol functions in the implementation or not. In fact, one could consider these options as Boolean parameters, which are assigned values in an implementation. We might call such parameters *IUT parameters*. Strictly speaking, *installation specific parameters* without much bearing on protocol logic, such as the net address of the installed IUT, also determine the behavior of an IUT. They may, for example, be the values of PDU fields produced by the IUT. Finally, *test specific parameters*, such as the net address of the lower tester and time limits not imposed by the specification, but introduced for the purpose of limiting the time spent for each test case, influence the behavior of the IUT and/or tester.

In all, OSI specifications usually define a full spectrum of many different concrete implemented behaviors, where the chosen parameter values and options describe the actual variant chosen for an IUT. Parameter values and other specific values for an IUT are specified in ICS and IXIT documents, cf. Subsections 2.4.1 and 2.4.2. IUT parameters are usually determined in the ICS, installation and test specific parameters are usually determined in the IXIT.

In the same sense, a test suite for a parameterized OSI specification cannot completely determine the tester's behavior. Rather, this behavior will be affected by the same parameters as the specification for which it is designed. It is only after their value assignment that a test suite will determine completely the behavior of the tester in all conceivable circumstances, including the creation of test data and the assignment of test verdicts.

Test suite parameters are declared as shown in Table 3.16. For each applicable parameter, they should identify clearly, but informally, the associated ICS or IXIT entry which provides its value. Parameters may become inapplicable, because certain ICS/IXIT entry values entail that other entries are not filled. We cannot tell for sure yet whether the title of the third column will change to "ICS/IXIT Ref". The degree of formality of these references and content values depends on the formalization of ICS and IXIT proformas.

A test suite parameter may be used in test case selection, and is expected to play the role of a constant during the execution of the test suite on a given IUT. Therefore, test suite parameters are assigned values only via their declarations and the corresponding ICS and IXIT values and not by means of assignments within test cases.

TABLE 3.16. Test suite parameter declarations

TEST SUITE PARAMETER DECLARATIONS			
PARAMETER NAME	TYPE	PICS/PIXIT REF	COMMENTS
:	:	:	:
TS_ParIdentifier	*Type* a predefined, ASN.1, test suite or PDU type	*FreeText* e.g. PICS question 1	[*FreeText*]
:	:	:	:
DETAILED COMMENTS: [*FreeText*]			

3.4.5 Test case selection

Test groups and test cases may be associated with test case selection expressions, Boolean expressions whose actual values determine whether the respective groups or cases are executed. This association is achieved in several steps: the test suite structure table and the test case index assign selection expression references to test groups and test cases, respectively (cf. 3.3.2). These expression identifiers are associated with actual expressions in the test case selection expression definitions table, whose pattern is shown in Table 3.17. These expressions will be evaluated to TRUE or FALSE on the basis of the ICS/IXIT entries. Test case selection expressions are used accumulatively: a test case is run only if

- its own selection expression in the test case index evaluates to TRUE, and
- all selection expressions of superior groups which contain it, as listed in the test suite structure, evaluate to TRUE.

A selection expression of TRUE is usually omitted, as this is the default value. Therefore, we may consider selection expressions as "deselection expressions."

TABLE 3.17. Test case selection expression definitions

TEST CASE SELECTION EXPRESSION DEFINITIONS		
EXPRESSION NAME	SELECTION EXPRESSION	COMMENTS
:	:	:
SelectExprIdentifier	*SelectionExpression* formed of literal values, TS parameters/ constants, other selection expressions	[*FreeText*]
:	:	:
DETAILED COMMENTS: [*FreeText*]		

3.4.6 Constants and variables

TEST SUITE CONSTANTS

Similarly to the values of test suite parameters, the values of test suite constants remain fixed during the execution of a test suite. The difference between the two lies in the point in time at which these values are fixed, as well as in the source of the values. Parameter values are obtained directly from ICS/IXIT entries, i.e. they are determined by the client or test laboratory prior to the execution of the test suite. Constant values, in contrast, are either directly assigned by the test suite writer, or they are calculated, according to the directions of the test suite writer. Thus, they are fixed either at the time of test suite writing or at the start of its execution. Their declarations are given collectively in a table following to the pattern shown in Table 3.18.

TABLE 3.18. Test suite constant declarations

TEST SUITE CONSTANT DECLARATIONS			
CONSTANT NAME	TYPE	VALUE	COMMENTS
:	:	:	:
TS_ConstIdentifier	*Type* not an ASP type	*DeclarationValue* an expression without TS or TC variables	[*FreeText*]
:	:	:	:
DETAILED COMMENTS: [*FreeText*]			

9646-3 says explicitly that TS constants are not derived from the (P)ICS or (P)IXIT, but permits TS parameters in the expressions defining their declared values. At any rate, we cannot imagine any problems with TS constant values derived from TS parameters.

TEST SUITE AND TEST CASE VARIABLES

As test suite and test case variables may also be assigned initial values, their declarations resemble very much those of test suite constants. They are given in two tables, whose patterns are shown in Tables 3.19 and 3.21.

TABLE 3.19. Test suite variable declarations

TEST SUITE VARIABLE DECLARATIONS			
VARIABLE NAME	TYPE	VALUE	COMMENTS
:	:	:	:
TS_VarIdentifier	*Type* not an ASP type	[*DeclarationValue*] an *expression* without TS or TC variables	[*FreeText*]
:	:	:	:
DETAILED COMMENTS: [*FreeText*]			

In using test suite variables, one should bear in mind that the order of the test cases within the ATS is not necessarily the order in which they are executed when the test suite is run. One cannot expect to save a value from one test case for a test case *listed later* in the test suite. But one can save a value for any test case which might happen to be *executed later*. Thus, TS variables can be used to save information from previously executed test cases, such as the last protocol state reached, as in the example given in Table 3.20.

TABLE 3.20. An example of a test suite variable

TEST SUITE VARIABLE DECLARATIONS			
VARIABLE NAME	TYPE	VALUE	COMMENTS
:	:	:	:
LastState	IA5STRING	"idle"	state of the lower tester at the end of the previous test case

In the majority of instances, test cases will be executed in the order in which they are listed in the ATS, simply because there will not be any reason to reorder them. ATS specifiers can state requirements on the order of execution (cf. 2.6.5), but TTCN does not provide any table for this purpose. Comments or an informal annex could be used instead.

In concurrent TTCN, only the main test component may use test suite variables.

TABLE 3.21. Test case variable declarations

TEST CASE VARIABLE DECLARATIONS			
VARIABLE NAME	TYPE	VALUE	COMMENTS
:	:	:	:
TC_VarIdentifier	*Type* not an ASP type	[*DeclarationValue*] an *expression* without TS or TC variables	[*FreeText*]
:	:	:	:
DETAILED COMMENTS: [*FreeText*]			

Note that with each new test case, a new instance of each test case variable is created, endowed with the initial value as declared in the value column, and with an undefined value if there is no corresponding entry. In concurrent TTCN (cf. Section 3.1), this instance belongs to the main test component. A newly started parallel test component gets an own instance of each test case variable, with the defined initial value (if any). Upon termination of a non-concurrent test case, the values of its test case variables return to undefined. Likewise, upon termination of a test component, the values of its instances of test case variables return to undefined.

3.4.7 PCOs and CPs

Points of control and observation (PCO) are declared as in Table 3.22. As described in Sections 2.5 and 3.1, their number and their positions relative to the IUT, the tester components and the underlying service provider depend upon the chosen test method.

Since PCOs need not coincide with single layer boundaries, the comment column may have to be used to complete the description of the PCO type, e.g. by giving the list of ASPs and PDUs that can occur at this PCO.

The MuxValue option, which was not contained in the first edition of TTCN, will be introduced by [ISO 9646-3-1]. It allows one to select a test suite parameter which identifies one connection (e.g. SAP, connection end point, logical channel id) out of the set of connections at the service boundary, to be handled by the PCO. Thus, test suite writers can leave multiplexing and demultiplexing to the means of testing, instead of having to specify these mechanisms explicitly in TTCN.

TABLE 3.22. PCO declarations

PCO DECLARATIONS			
PCO NAME	PCO TYPE	ROLE	COMMENTS
: *PCO_Identifier* :	: *PCO_TypeIdentifier* ["("*MuxValue*")"] e.g. TSAP or NSAP(muxparm1) :	: ***UT* | *LT*** :	: [*FreeText*] :
DETAILED COMMENTS: [*FreeText*]			

In contrast to PCOs, the declarations of coordination points (cf. 2.5.6 and 3.1) do not give any information other than their names.

TABLE 3.23. CP declarations

COORDINATION POINT DECLARATIONS	
CP NAME	COMMENTS
: *CP_Identifier* :	: [*FreeText*] :
DETAILED COMMENTS: [*FreeText*]	

3.4.8 Timers

Test suites may make use of *timers*, which act like stop watches. Timers have to be declared, as in Table 3.24. In their declaration, they always are assigned an unalterable time unit, and they may be given a positive integer default setting (a duration, counted in declared units). Such a default setting may be overwritten in test case behavior descriptions in the dynamic part of the test suite. The operational semantics of timers are described in Chapter 4.

TABLE 3.24. Timer declarations

TIMER DECLARATIONS			
TIMER NAME	DURATION	UNIT	COMMENTS
:	:	:	:
TimerIdentifier	[*DeclarationValue*] an integer valued *expression* without variables	*TimeUnit* i.e. ***ps*** | ***ns*** | ***us*** | ***ms*** | ***s*** | ***min***, where "us" means "μs"	[*FreeText*]
:	:	:	:
DETAILED COMMENTS: [*FreeText*]			

3.4.9 Test components and their configurations

In concurrent TTCN, test components and test component configurations are used. Tables in the declarations part are designated for all test components collectively, and for each test component configuration separately. The declaration of test components follows the pattern shown in Table 3.25. Note that only the "quantitative connectivity" of the test components is defined, i.e. the numbers of the PCOs and CPs used, but not the names of these interaction points, let alone their (PCO) types. Moreover, each test component is assigned a single role as either a main or a parallel test component.

TABLE 3.25. Test component declarations

TEST COMPONENT DECLARATIONS				
COMPONENT NAME	COMPONENT ROLE	NR PCOS	NR CPS	COMMENTS
:	:	:	:	:
TCompIdentifier	***MTC*** | ***PTC***	*Num_PCOs* a *number*, i.e. an integer ≥0	*Num_CPs* a *number*, i.e. an integer ≥0	[*FreeText*]
:	:	:	:	:
DETAILED COMMENTS: [*FreeText*]				

In a test component configuration, a number of test components, more precisely one MTC and zero or more PTCs, are symbolically interconnected and inserted into the test architecture: their pairwise interconnectivity is defined by associating CPs with them, each CP connecting exactly two test components.

Their external connectivity is defined by naming the PCOs that each test component uses. The association of CPs and PCOs with a test component must, of course, agree with the number of CPs and PCOs declared for it. The pattern for a test component configuration declaration is shown in Table 3.26.

TABLE 3.26. Test component configuration declaration

<table>
<tr><td colspan="4">TEST COMPONENT CONFIGURATION DECLARATION</td></tr>
<tr><td colspan="4">CONFIGURATION NAME : *TCompConfigIdentifier*
COMMENT : [*FreeText*]</td></tr>
<tr><td>COMPONENTS USED</td><td>PCOS USED</td><td>CPS USED</td><td>COMMENTS</td></tr>
<tr><td>:
TCompIdentifier
:</td><td>:
[*PCO_List*]
e.g. PCO_A, PCO_B
:</td><td>:
[*CP_List*]
e.g. CP_1, CP_2
:</td><td>:
[*FreeText*]
:</td></tr>
<tr><td colspan="4">DETAILED COMMENTS: [*FreeText*]</td></tr>
</table>

3.4.10 ASP types

The types of ASPs expected to be sent and received at PCOs are declared in the declarations part of the test suite. Note that ASPs providing a two-way exchange of data between user and provider are not really expected in TTCN, cf. 4.9.1. Like general test suite types, ASP types may be declared in a purely tabular manner, by ASN.1 or by ASN.1 reference. In each case, the table header gives the ASP type name, indicates at which PCO types ASPs of this type can occur, and optionally includes a comment.

A purely tabular ASP type definition has very much in common with a structured type definition. The top level components of the ASP are not called elements, but *parameters*. ASP parameters are implicitly considered as optional. Their names are given either as in the relevant service standard (if syntactically acceptable as TTCN identifiers) or in a short form, followed by the standardized name in parentheses. In the latter case, the short name may be used in the dynamic part. The same alternatives exist for the ASP (type) name. Table 3.27 shows the pattern of an ASP type definition.

Length attributes may be used to further restrict the types used, e.g. if the parameter corresponds to a PDU field. Length attributes are addressed in 3.4.2. Two other language elements in this table pattern need explanation: the macro symbol (“<-”) and the PDU pseudotype.

TABLE 3.27. ASP type definition

ASP TYPE DEFINITION		
ASP NAME : *ASP_Identifier* ["(" *FreeText* ")"] PCO TYPE : [*PCO_TypeIdentifier*] COMMENTS : [*FreeText*]		
PARAMETER NAME	PARAMETER TYPE	COMMENTS
: *ASP_ParIdentifier* ["(" *FreeText* ")"] | "<-" :	: (*Type* [*LengthAttribute*]) | ***PDU*** :	: [*FreeText*] :
DETAILED COMMENTS: [*FreeText*]		

The *macro symbol* in the position of a parameter name, followed by the identifier of a structured type in the parameter type column, means that this row in the table should be replaced (in mind) by the rows of the body of the structured type declaration. Thus, the designated sequence of element type declarations of another structured data type (see 3.4.2) is, in a way, textually inserted in this place, between and in the same rank with the other parameter type information – hence the name "macro."

Let us consider an example: Table 3.28 describes a structured type Address, which is then used as a macro in Table 3.29 to define ASP_1. If the auxiliary type is not used anywhere else, both tables could be replaced by one: Table 3.30. Note that ASP_1 is a different type than ASP_2, which is declared in Table 3.31: ASP_1 has three parameters, ASP_2 only two. One of ASP_2's parameters is structured, thus giving ASP_2 a greater "structural depth" than ASP.1.

TABLE 3.28. An auxiliary structured type example

STRUCTURED TYPE DEFINITION		
TYPE NAME : Address COMMENTS : This type will be used to define ASP_1.		
ELEMENT NAME	TYPE DEFINITION	COMMENTS
SA (Sender Address)	OCTETSTRING[2]	
RA (Receiver Address)	OCTETSTRING[2]	

TABLE 3.29: An ASP type definition with macro

ASP TYPE DEFINITION		
ASP NAME : ASP_1 PCO TYPE : LowerTesterPCO COMMENTS : an example of macro use		
PARAMETER NAME	PARAMETER TYPE	COMMENTS
<- Contents	Address OCTETSTRING[252]	... represents 2 parameters

TABLE 3.30: An equivalent ASP type definition without macro

ASP TYPE DEFINITION		
ASP NAME : ASP_1 PCO TYPE : LowerTesterPCO COMMENTS : an example with only "explicit" parameters		
PARAMETER NAME	PARAMETER TYPE	COMMENTS
SA (Sender Address) RA (Receiver Address) Contents	OCTETSTRING[2] OCTETSTRING[2] OCTETSTRING[252]	first Address parameter second Address parameter

TABLE 3.31. A related – but not equivalent – ASP type definition

ASP TYPE DEFINITION		
ASP NAME : ASP_2 PCO TYPE : LowerTesterPCO COMMENTS : an example with a structured parameter		
PARAMETER NAME	PARAMETER TYPE	COMMENTS
FromTo Contents	Address OCTETSTRING[252]	one structured parameter

The pseudotype "PDU" in the parameter type column permits to refer from ASP constraints of the given ASP type to PDU constraints of any PDU type in the constraints part, cf. the note after Table 3.10. Restrictions about components of type "PDU" will be stated in 4.3.1. PDU types are introduced in the next subsection, and the constraint reference technique called chaining is described in Section 3.5.

The pattern for an ASN.1 ASP type definition is shown in Table 3.32. It closely resembles the pattern for ASN.1 type definitions in Table 3.12: its single column contains an entry of the form *ASN1_Type&LocalTypes,* and the same rules apply with respect to the full and short type name, the visibility of local auxiliary types, and the use of dashes and underscores. The macro operator "<-" and the pseudo-type "PDU" do not occur in ASN.1.

TABLE 3.32. ASN.1 ASP type definition

ASN.1 ASP TYPE DEFINITION
ASP NAME : *ASP_Identifier* ["(" *FreeText* ")"] PCO_TYPE : [*PCO_TypeIdentifier*] – may be omitted if only one PCO is involved COMMENTS : [*FreeText*]
TYPE DEFINITION
ASN1_Type&LocalTypes
DETAILED COMMENTS: [*FreeText*]

The pattern for an ASN.1 ASP type definition by reference is shown in Table 3.33. The major difference from the corresponding general type definition in Table 3.14 is the additional PCO type column. The remarks made about that table apply here, too.

TABLE 3.33. ASN.1 ASP type definition by reference

ASN.1 ASP TYPE DEFINITIONS BY REFERENCE				
ASP NAME	PCO TYPE	TYPE REFERENCE	MODULE IDENTIFIER	COMMENTS
: *ASP_Identifier* ["(" *FreeText* ")"] :	: [*PCO_TypeIdentifier*] necessary if ≥ 1 PCO :	: *TypeReference* an ASN.1 "typereference" :	: *ModuleIdentifier* in the form defined in ASN.1 :	: [*FreeText*] :
DETAILED COMMENTS: [*FreeText*]				

3.4.11 PDU types

The next group of type definitions deals with the types of PDUs expected to be sent and received at PCOs. Like general test suite types and ASP types, PDU types may be declared in a purely tabular manner, by ASN.1 or by ASN.1 reference. In each case, the table pattern very much resembles that of structured and ASP types, as can be seen by comparing Tables 3.34, 3.35 and 3.36 with the corresponding earlier tables. The top level components of PDUs, which correspond to the elements of structured types and the parameters of ASPs, are called *fields*. PDU type definitions are subject to the same conventions. For example, fields are implicitly considered to be optional, and in the purely tabular definition their names – just like the PDU names themselves – must either agree with the standardized field (or PDU) names or refer to them in parentheses.

TABLE 3.34. PDU type definition

PDU TYPE DEFINITION		
PDU NAME : *PDU_Identifier* ["(" *FreeText* ")"] PCO TYPE : [*PCO_TypeIdentifier*] COMMENTS : [*FreeText*]		
FIELD NAME	FIELD TYPE	COMMENTS
: *PDU_FieldIdentifier* ["(" *FreeText* ")"] | "<-" :	: (*Type*[*LengthAttribute*]) | ***PDU*** :	: [*FreeText*] :
DETAILED COMMENTS: [*FreeText*]		

TABLE 3.35. ASN.1 PDU type definition

ASN.1 PDU TYPE DEFINITION
TYPE NAME : *PDU_Identifier* ["(" *FreeText* ")"] PCO_TYPE : [*PCO_TypeIdentifier*] – may be omitted if only one PCO is involved COMMENTS : [*FreeText*]
TYPE DEFINITION
ASN1_Type&LocalTypes
DETAILED COMMENTS: [*FreeText*]

TABLE 3.36. ASN.1 PDU type definition by reference

ASN.1 PDU TYPE DEFINITIONS BY REFERENCE				
PDU NAME	PCO TYPE	TYPE REFERENCE	MODULE IDENTIFIER	COMMENTS
:	:	:	:	:
PDU_Identifier ["("*FreeText*")"]	[*PCO_TypeIdentifier*] necessary if ≥ 1 PCO	*TypeReference* an ASN.1 "typereference"	*ModuleIdentifier* in the form defined in ASN.1	[*FreeText*]
:	:	:	:	:
DETAILED COMMENTS: [*FreeText*]				

3.4.12 Encoding information

In some protocols, PDUs may be encoded in various ways. Let us call each method of encoding and decoding PDUs an *encoding rule set*. Some encoding rule sets support alternatives, which we call *encoding variations*. If various encoding rules or encoding variations are applicable to a protocol, the goal of making tester behavior deterministic cannot be achieved, unless the test suite predetermines the rule and variation used in every single SEND event. Moreover, test purposes may call for the investigation of proper encoding and decoding by the IUT. TTCN Amendment 2 [ISO 9646-3-2] provides language elements to handle standardized encoding rule sets. All encoding rule sets considered in an ATS and identified by a reference to standard, or a clause of a standard, are named with TTCN identifiers in an encoding definitions table, cf. Table 3.37.

TABLE 3.37. Encoding definitions

ENCODING DEFINITIONS			
ENCODING RULE NAME	REFERENCE	DEFAULT	COMMENTS
:	:	:	:
EncodingRuleIdentifier the name by which the encoding rule set will be referenced	*EncodingReference* a FreeText reference to a (clause of a) standard defining the encoding rule set	[*DefaultExpression*] giving a Boolean value	[*FreeText*]
:	:	:	:
DETAILED COMMENTS: [*FreeText*]			

Encoding definitions only apply to PDU constraints. To provide for the automatic selection of an encoding rule set, Boolean expressions, called default expressions, are associated with each encoding. At the moment during test case execution when the right encoding has to be chosen, at most one of the default expressions must evaluate to TRUE. It has been suggested that the first encoding is selected if none of the expressions yields TRUE, even though this rule may mask test specification errors.

For each encoding rule set, a similar table may declare encoding variations, as shown in Table 3.38. Encoding variations apply to PDUs, structured and ASN.1 types, and PDU fields.

TABLE 3.38. Encoding variations

ENCODING VARIATIONS			
ENCODING RULE NAME : *EncodingRuleIdentifier* TYPE LIST : [*TypeList*] – zero or more names of defined types, separated by commas. An empty list is interpreted as the list of all types used for fields in the PDU. COMMENT : [*FreeText*]			
ENCODING VARIATION	REFERENCE	DEFAULT	COMMENTS
:	:	:	:
EncVariationIdentifier the name by which the encoding variation will be referenced	*VariationReference* a FreeText reference to a (clause of a) standard defining the encoding variation	[*DefaultExpression*] giving a Boolean value	[*FreeText*]
:	:	:	:
DETAILED COMMENTS: [*FreeText*]			

Unfortunately, the granularity of encoding rule standards does not always permit determination of encoding entirely by referencing a clause. The Boolean value TRUE, for example, is permitted in clause 8.2 of [ISO 8825] to be encoded as any of the 255 bitstrings between '00000001 00000001 00000001'B and '00000001 00000001 11111111'B. In such cases, the comments column will have to be used to describe informally the encoding actually meant.

In thorough testing, it is desirable to investigate the IUT's reactions to invalid encodings. For this purpose, invalid encodings may be defined for any type attributed to PDU fields in the test suite, as shown in Table 3.39.

TABLE 3.39. Invalid field encoding definition

INVALID FIELD ENCODING OPERATION DEFINITION		
INVALID FIELD ENCODING NAME	:	*InvalidFieldEncoding&ParList* – an identifier, optionally followed by a list of formal parameters
TYPE LIST	:	[*TypeList*] – as above
COMMENTS	:	[*FreeText*]
ENCODING OPERATION DEFINITION		
[*VarBlock*] *ProcStatement* or *FreeText* – i.e. either a formal or an informal definition of the operation, as in the case of test suite operations, see 4.3.3. The operation must define a mapping of any of the listed types to bitstrings.		
DETAILED COMMENTS: [*FreeText*]		

The optional parameters for such a field encoding operation do not represent the field values to be encoded; rather, they can be used to control the way a field value is encoded. Amendment 2 contains an elaborate priority scheme explaining for each mode of encoding specification and each application level which encoding definitions are possible and which override others. The highest priority is assigned to invalid field encoding definitions.

3.4.13 CM types

This last group of type definitions in the declaration part of an ATS deals with the coordination messages to be exchanged between test components at CPs. In order to grant a maximum of liberty to the test suite writer, they can be built up in arbitrary complexity from arbitrary types, just as ASPs and PDUs. As CMs are TTCN and testing specific, it is not assumed that other standards will provide names or types for CMs. Therefore, their names need not relate to standardized names, and there is no provision for ASN.1 definitions by reference. The reference to particular PCO types is also dropped, and it is not replaced by a list of CP names. Apart from that, and taking into account the fact that the components of CMs are called parameters, the CM type definition table pattern, shown in Figure 3.40, and the ASN.1 CM type definition table pattern, shown in Figure 3.41, correspond to the respective patterns for ASPs and PDUs, which are shown in Tables 3.27, 3.32, 3.34, and 3.35.

TABLE 3.40. CM type definition

CM TYPE DEFINITION		
CM NAME : *CM_Identifier* COMMENTS : [*FreeText*]		
PARAMETER NAME	PARAMETER TYPE	COMMENTS
: *CM_ParIdentifier* | "<-" :	: (*Type* [*LengthAttribute*]) | ***PDU*** :	: [*FreeText*] :
DETAILED COMMENTS: [*FreeText*]		

TABLE 3.41. ASN.1 CM type definition

ASN.1 CM TYPE DEFINITION
CM NAME : *CM_Identifier* COMMENTS : [*FreeText*]
TYPE DEFINITION
ASN1_Type&LocalTypes
DETAILED COMMENTS: [*FreeText*]

3.4.14 Renaming and abbreviations

In ASP and PDU type definitions, we encountered the possibility of choosing between the standardized names and specifically assigned names for ASPs and PDUs. Let us call the chosen names (standardized or new) "type names" for the moment. Moreover, TTCN provides the possibility of renaming ASP and PDU type names in a separate table, as shown in Table 3.42. Let us call the alias names assigned this way "type aliases." Type aliases may only be used in the dynamic part. Type names have to be used when the ASP/PDU types are referenced from other type definitions or from constraint declarations. In the behavior descriptions of the dynamic part, both type names and type aliases are allowed.

In practice, alias names are used both for abbreviating long type names and for indicating the particular use of an ASP or PDU. An example of the latter use is calling a DATArequest ASP a CR, if it contains a CR-PDU.

Two levels of renaming (standardized name → type name → type alias) should only be used used for good reasons, and should be accompanied by appropriate comments. Otherwise, double renaming will lead to confusion. So will mixed use of type names and aliases.

By the way, *AliasIdentifiers* do not appear on the right side of any BNF rule, apart from those for the alias definitions, because aliasing and its resolution are considered to be separate measures that should not be reflected in BNF. This special treatment seems to be a relic from early TTCN versions, in which arbitrary pieces of TTCN.MP string could be renamed.

TABLE 3.42. Alias definitions

ALIAS DEFINITIONS		
ALIAS NAME	EXPANSION	COMMENTS
:	:	:
AliasIdentifier	*Expansion*	[*FreeText*]
:	:	:
DETAILED COMMENTS: [*FreeText*]		

3.4.15 ICS/IXIT and declared values

Throughout several subsections we have observed the possible ways by which data can flow from other test documents into an ATS written in TTCN. Values from ICS and IXIT entries are obtained by the test suite via test suite parameters. There, when used in expressions, they become available for the determination of initial or current values of constants and variables. Figure 3.5 gives an overview of the valid paths of information flow towards constants and variables in the declarations part.

3.5 Constraints

In TTCN, data object values or patterns for data objects are called *constraints*. In fact, constraints can always be considered as patterns, some of which are simply matched by only one data object. The behavior descriptions in the dynamic part of a test suite do not spell out the entire constraint, i.e. the full description

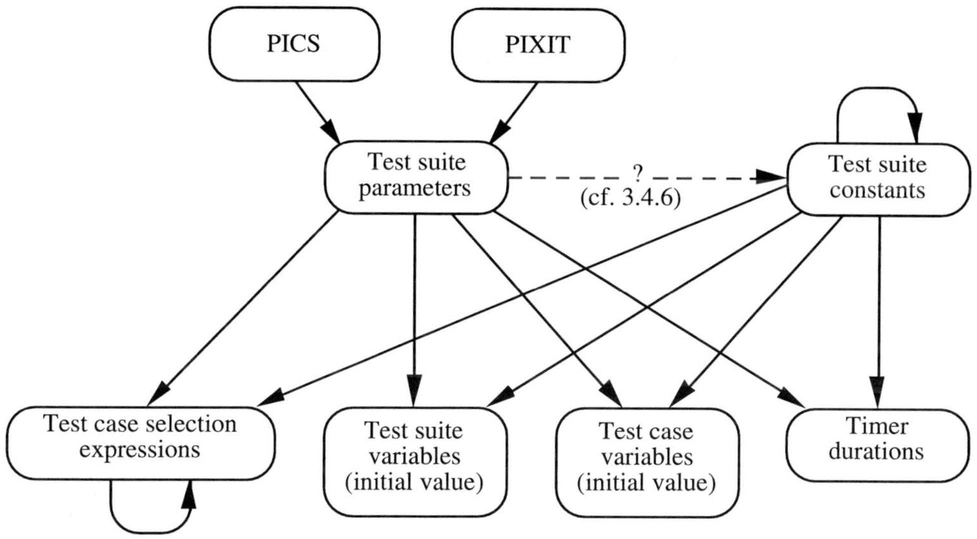

FIGURE 3.5. Data flow to constants and variables in the declarations part

of the ASPs, PDUs, and CMs involved in the send and receive actions of the tester. Integrated constraint specifications would clutter test case behavior descriptions and make it hard to perceive the event dynamics. Hence, constraints are only referenced by their names, which are usually expressive acronyms and abbreviations. Their specifications are provided in a separate part of the test suite, the constraints part. A benefit of this approach is that identical constraints used in several places in the dynamic part need only be described once.

3.5.1 General principles

In this subsection we discuss the general properties of constraints, their specification and their use. We deal with the relationship between constraints and types, the use of variables and expressions in constraints, with parameterization, "chaining," and the use of constraints in send and receive events. To avoid tiring repetitions, we will use the non-standard term *component* as a generic term for ASP and CM parameters, PDU fields, and structured type elements.

To support modularity, constraints can be specified not only for ASPs, PDUs and CMs, but also for auxiliary structured and ASN.1 types. Each constraint is declared to be of a given type, and its entire declaration is closely associated with the definition of this type. Constraints can be declared either in a purely tabular or in an ASN.1 oriented manner. They have the same component structure and component types as the types assigned to them. Constraint declarations assign values and value patterns to the components of the constraints by means of literal values, data object references, expressions, "matching mechanisms," and constraint references.

Constraint specifications permit a flexible description of data patterns by means of a multitude of language elements for pattern definition, called *matching mechanisms*. Thus, for example, constraint components may be defined as value ranges, or with the aid of "wildcards."

Each send statement of the test suite contains a reference to a constraint which completely specifies an ASP, PDU or CM to be sent, and each receive statement references a constraint specifying the pattern of all ASPs, PDUs or CMs which are to be matched by this statement. Note that incoming data objects which do not match this pattern are not lost or discarded, but are treated by alternative receive statements or by an "otherwise" statement. More details on sends and receives are provided below and in Section 4.4.

Comparing a received ASP, PDU or CM against the pattern defined by a given constraint results only in a binary decision, "match" or "no match." This does not automatically cause the storing of the received data object or any of its components. Instead, value assignment in a receive event can be achieved only by explicit assignments (cf. 4.4.4). In fact, some experts would have preferred to be able to save values from received data objects as an implicit result of constraint matching. In the end, however, it was decided to make these value assignments explicit, to avoid unexpected side effects.

Any expression appearing in a constraint has to have a defined value when the constraint is used. Expressions in constraints may be built up from literal values, parameters, and constants, but they must not contain test case or test suite variables. The value of a variable may be used in a constraint, but only if passed as an actual parameter to this constraint. In such instances, the variable must already be bound to a value. This value is not changed when used by the constraint: constraint parameters are passed by value.

PARAMETERIZATION AND CHAINING OF CONSTRAINTS

As indicated above, constraints may be parameterized. In this case, the constraint name is followed by a list of formal parameters when the constraint is specified and by a list of actual parameters when the constraint is referenced. The formal parameters may appear in expressions assigned to constraint components. The general syntax of a formal parameter list is described in Subsection 3.2.4.

Constraint specifications may refer to other constraints, such that the "bigger" constraint is assembled from components, one of which is a "subconstraint." Referencing between constraints is called *chaining,* and is not permitted recursively. Chaining occurs if the value of a component of one constraint is:

- an explicit, and hence fixed, reference to another constraint of the appropriate type. This is called *static chaining.* For example, in a constraint definition for a network DATA request ASP, the value of the user data field may be a reference to a constraint of type CR-TPDU (transport connect request PDU).
- a formal parameter of the constraint, and in its parameterized reference in the dynamic part the corresponding actual parameter refers to another constraint. In this case, called *dynamic chaining*, the component is dynamically passed a constraint reference as an actual value. For example, in a constraint for a network DATA request ASP, the value of the user data field may be a formal parameter. This constraint may be referenced in the dynamic part with a reference to a constraint of type CR-TPDU, as the actual value of the formal parameter.

CONSTRAINTS IN SEND STATEMENTS

A constraint referenced in a send statement has to evaluate to a specific data object of the appropriate type. Evaluation to a unique value can be guaranteed by the constraint itself, or it can be achieved with the aid of assignments on the send statement, assigning specific values to the components which previously permitted more than one value. This is made possible by the rule that such "last minute assignments" override values declared for the constraint components, cf. 4.4.4. As a simple illustration, if a send statement references an ASP of type INTEGER, defined by the constraint *Constr*, it would be valid that

- *Constr* is declared to have a specific value, or
- *Constr* is declared to have the formal parameter *formalpar* as its value, and the actual value of *formalpar* is a specific value, or
- *Constr* is declared to permit an arbitrary integer, and an assignment on the send statement assigns a specific value.

In the following situations the use of *Constr* in the send would be invalid:

- *Constr* is declared to permit a range of values, and there is no assignment in the send to assign a specific value, or
- *Constr* is declared to have the value *formalpar*, and the actual value of *formalpar* is a constraint permitting a range of values.

Last minute assignments to constraints permit considerable savings in the constraints part if many minor modifications of one large constraint are needed. They prevent having to declare all of the variations in full detail. However, another feature of the constraints part provides a similar kind of versatility: the mechanism of *constraint modification*. The latter way, however, requires that for each constraint variant a separate, though shorter, table and an individual name be specified. The provision for constraint modification both by means of declarations and by means of send assignments seems to be a compromise between two styles of early TTCN use. It will help readability not to use both mechanisms in unrestrained combinations.

Although the standard is not perfectly clear about this point, it seems that the only non-unique patterns in constraints referenced in send statements – which consequently must be overwritten with specific values by send assignments – are AnyValue (?) and AnyOrOmit (*). At first sight it seems impossible to specify in send assignments "Omit this parameter," because what they assign may be a *Value* but not a *ConstraintValue*. Closer scrutiny reveals that a *Value* may be any *ConstraintValue&Attributes*. This follows from the BNF and static semantics rules:

Value ::= *LiteralValue* | *ASN1_Value* (rule in TTCN BNF)
ASN1_Value (TTCN BNF) ::= *Value* (ASN.1 BNF) (note in TTCN BNF)
Value ::= *BuiltinValue* | *DefinedValue* (rule in ASN.1 BNF)
DefinedValue (ASN.1 BNF) ::= *ConstraintValue&Attributes* (TTCN BNF)
(note in TTCN BNF)

On the one hand, these BNF rules allow to specify, for example, "user_data := OMIT" on a send statement, which is desirable. On the other hand, they syntactically allow lots of semantically undesired expansions, which then have to be ruled out by a number of static semantic restrictions. The authors are not aware whether this unrestricted loophole from values to constraint values was intentional. Alternative approaches offer clearer rules: it would simplify things to combine types and constraints, which describe constant or parameterized sets of values, into one single notion, well separated from the notions of constants and variables, which describe single values. In send events, variables might be used instead of constraints. Since the omission of an optional component does not create a set of possible values, it could be permitted explicitly in specific values of variable components, similarly as in ASN.1; but the other pattern-forming mechanisms should not.

As previously mentioned, components can be declared as *optional* either implicitly, in tabular types, or explicitly (as OPTIONAL), in ASN.1-defined types. The omission of an optional ASP or CM parameter or PDU field in a tabular and ASN.1 constraint can be expressed by using the explicit Omit for constraints, in ASN.1 also by plain omission of the parameter or field.

In order to maintain the desired determinism of tester behavior, component values of the ASN.1 types SET and SET OF are sent in the order in which they are listed in the constraint.

CONSTRAINTS IN RECEIVE STATEMENTS

A constraint referenced by a receive statement often cannot be defined as a unique data object, even if it uses dynamically determined information in the form of variables as actual parameters. The reason for this is that the ASP or PDU may contain components for which the IUT or the underlying provider supply a new value out of a choice of valid possibilities. The matching mechanisms permit the definition of choices in many ways.

A receive statement is executed only if the received ASP, PDU or CM matches the referenced constraint. The condition for matching is that all components of the received data object are consistent in their types and values with the respective types, values and patterns declared for the constraint. Generally, in order to match, components which are specified in the constraint must not be absent in the ASP, PDU or CM. However, if the type definition declares a component as optional, and if the component is specified in the constraint by one of the optionality related matching mechanisms (cf. Table 3.43), the component may be absent. In the case of structured components this consistency must apply recursively to all subcomponents. The decision of whether or not to execute the receive event, however, may depend additionally on a Boolean condition specified on the receive statement, cf. 4.4.3. PDUs are compared as decoded according to the specified encoding rules (cf. Subsection 3.4.12); it is irrelevant whether they would match the constraint if other encoding rules were applied.

Note that the consistency of a value with a type is not defined formally in the standard.

3.5.2 Data pattern descriptions

TTCN offers a powerful set of *matching mechanisms* that can be used to describe data object patterns. Syntactically they are used as substitutes, particles or add-ons for single values in specific data object descriptions; semantically they are used to define subsets of the relevant types. TTCN matching mechanisms can be grouped in three different ways:

- according to their *syntactic relation* to single specific values of a given type, i.e. whether
  - it actually is a specific value,
  - it is a substitute for a specific value,
  - it is used as a component of a value of a set or "sequential" type, or
  - it is used as an attribute to a value,
- according to their *semantic* way of subsetting the type, i.e. whether it represents
  - a purely set-theoretic operation possible on any type,

  or it is based on
  - the natural total order of the type INTEGER,
  - set properties of the data objects, or
  - sequence properties of the data objects , or
  - set or sequence properties of the data objects , or
  - the optionality of data objects, and, finally,
- according to the *types* to which they can be applied.

Table 3.43 lists the matching mechanisms, grouped according to the first two aspects. Table 3.44 summarizes the applicability of matching mechanisms to specific data types. The types above the middle line are predefined types common to ASN.1 and TTCN; the types below are exclusively ASN.1 type constructors. "Enumerated" is meant to embrace both the TTCN tabular types defined as *SimpleValueList* and the ASN.1 types defined as *EnumeratedType*.

For each matching mechanism, Table 3.45 gives one or more typical examples of matching and non-matching values. "—" signifies "no value".

MATCHING MECHANISMS FOR ARBITRARY TYPES

A simple pattern for values of a given arbitrary type is a *specific value*, defined by an expression yielding a value of that type. A value of a tabular ASP, PDU or structured type, or of an ASN.1 SEQUENCE or SEQUENCE OF type, matches a given specific value if the latter is of the respective type and the components of both values match in the given order. SET and SET OF values may be in any order.

TABLE 3.43. A typology of TTCN matching mechanisms

	SPECIFIC VALUE	SUBSTITUTE	COMPONENT	ATTRIBUTE
SET-THEORETIC ON TYPE	Specific value	ValueList Complement AnyValue		
TOTAL ORDER ON TYPE		ValueRange		
SET STRUCTURE		Superset Subset		
SEQUENCE STRUCTURE			Permutation	
SET/SEQUENCE STRUCTURE			AnyOne AnyOrNone	Length
RELATED TO OPTIONALITY		Omit AnyOrOmit		IfPresent

TABLE 3.44. The applicability of TTCN matching mechanisms

MATCHING MECHANISM / TYPE	SpecificValue	ValueList	Complement	AnyValue (?)	ValueRange	Superset	Subset	Permutation	AnyOne (?)	AnyOrNone (*)	Length	Omit (-)	AnyOrOmit (*)	IfPresent
BOOLEAN	•	•	•	•								•	•	•
INTEGER	•	•	•	•	•							•	•	•
ENUMERATED	•	•	•	•								•	•	•
BITSTRING	•	•	•	•					•	•	•	•	•	•
OCTETSTRING	•	•	•	•					•	•	•	•	•	•
HEXSTRING	•	•	•	•					•	•	•	•	•	•
CHARSTRINGS	•	•	•	•					•	•	•	•	•	•
SEQUENCE	•	•	•	•								•	•	•
SEQUENCE OF	•	•	•	•				•	•	•	•	•	•	•
SET	•	•	•	•								•	•	•
SET OF	•	•	•	•		•	•		•	•	•	•	•	•
ANY	•	•	•	•								•	•	•
CHOICE	•	•	•	∘								•	•	•
OBJECT ID	•	•	•	•								•	•	•

TABLE 3.45. Examples of the application of TTCN matching mechanisms

MATCHING MECHANISM	TYPE	CONSTRAINT VALUE	MATCHING VALUES	NON-MATCHING VALUES
Specific value	INTEGER BOOLEAN	5 TRUE	5 TRUE	1 FALSE
ValueList	INTEGER	(1,10,100,1000)	10	2, TRUE
Complement	INTEGER BOOLEAN BITSTRING[2]	COMPLEMENT(1) COMPLEMENT (TRUE) COMPLEMENT('11'B)	1000 FALSE '10'B, '00'B	1 TRUE '11'B, '1'B
AnyValue	INTEGER(0..10) BOOLEAN BITSTRING[2]	? ? ?	5 TRUE '11'B	20, — FALSE, 5 '111'B
ValueRange	INTEGER INTEGER	(10..20) (0 .. INFINITY)	15 10	21, '1'B -10
SuperSet	SET OF INTEGER	SUPERSET({1,2,3})	{10,1,20,2,3}	{1,2,10,20}
SubSet	SET OF INTEGER	SUBSET({10,20,30})	{10,20}	{10,100}
Permutation	SEQUENCE OF INTEGER SEQUENCE OF INTE GER	{PERMUTATION (1,2,3)} {PERMUTATION (1,2,*)}	(1,3,2) (3,4,2,1)	(1,2,3,4) (3,4)
AnyOne	IA5STRING IA5STRING	"A?C" "ab\\c\?"	"ABC" "ab\c?"	"AB","ABB" "ab\\c\?"
AnyOrNone	IA5STRING BITSTRING	"A*Z" '1111'*B[4..8]	"ABCZ","AZ" '11110000'B	"aBCDZ" '111100001'B
Length	BITSTRING IA5STRING	'11*00'B[4..6] "ABC*ABC" [8]	'111000'B, '1100'B "ABC12ABC"	'1111000'B, '111000'H "ABC1ABC"
Omit	BITSTRING[2] OPTIONAL	-, OMIT	—	'00'B, '1111'B
AnyOrOmit	BITSTRING[2] OPTIONAL IA5STRING OPTIONAL	* *	—, '11'B —, "ABC"	'1111'B '0000'B
If_Present	BITSTRING[1] OPTIONAL BITSTRING[1] OPTIONAL	'1'B IF_PRESENT ? IF_PRESENT	—, '1'B '0'B, '1'B	'0'B, FALSE '00'B

ValueList permits any value of a given arbitrary type which is matched by at least one of those in a finite list of constraint values.

*ValueList** ::= "(" *ConstraintValue&Attributes* { ","
ConstraintValue&Attributes } ")"

Examples, as for all other matching mechanisms, can be found in Table 3.45.

Complement permits the values excluded by ValueList: it permits any value of a given arbitrary type unless the value matches any one of those in a finite list of constraint values.

Complement ::= ***COMPLEMENT*** *ValueList*

The pattern that matches any single value of a given arbitrary type is called *AnyValue* and is written as a question mark.

AnyValue ::= "?"

A MATCHING MECHANISM FOR INTEGERS

A *ValueRange* may be given as a constraint value of type INTEGER. The syntax of ValueRange specifications in constraints resembles that of their IntegerRange counterparts used with types, as explained at the end of Subsection 3.4.2. Range boundaries may now be constraint expressions which have to evaluate to specific INTEGER values. This appears to be very much like requiring expressions, in which case integer ranges in types and in constraints would really be identical.

ValueRange::="(" *ValRange* ")"
*ValRange** ::= (*LowerRangeBound* (***TO***|"..") *UpperRangeBound*)
*LowerRangeBound** ::= *ConstraintExpression* | "-" ***INFINITY***
UpperRangeBound ::= *ConstraintExpression* | ***INFINITY***

MATCHING MECHANISMS FOR SETS

SuperSet may be used for a constraint value of the ASN.1 SET OF type. A value of the corresponding type matches a superset expression SUPERSET(x) if it is a superset of a set matching the constraint value x. Note that x may be a complex constraint value matching many diffent sets. In the simplest case, where x evaluates to a specific set, it is exactly the supersets of that set (relative to the given type) which match. It is recommended that use of SuperSet be confined to this simple case.

SuperSet ::= ***SUPERSET*** "(" *ConstraintValue&Attributes* ")"

SubSet, too, may be used for a constraint value of the ASN.1 SET OF type. A value of the corresponding type matches a subset expression SUBSET(x) if it is a subset of a set matching the constraint value x. The simpler cases of SubSet correspond to those of SuperSet.

SubSet ::= ***SUBSET*** "(" *ConstraintValue&Attributes* ")"

MATCHING MECHANISMS FOR SEQUENCES AND STRINGS

Permutation may be placed, like a single of several elements, inside the braces enclosing an ASN.1 SEQUENCE_OF value. It permits all the elements of its argument list to appear in arbitrary order between the previous and the subsequent values in the sequence. Thus, {1,PERMUTATION(2,3)} is equivalent to the value list ({1,2,3},{1,3,2}).

Permutation ::= ***PERMUTATION*** *ValueList*

If AnyOrNone (cf. below) is one of the arguments of a Permutation term, the former is considered to be evaluated first. Thus, {PERMUTATION(a,*)} means "an arbitrary sequence containing at least one a" and does not mean "a, followed by an arbitrary list of components," cf. the Permutation examples in Table 3.45.

AnyOne may be used in a literal constraint value of either a string, a SEQUENCE_OF or a SET OF type. It is written as a question mark. Any single character, sequence element or set member, respectively, matches this pattern.

AnyOne ::= "?"

To prevent ambiguity, any character "?" in a character string must be represented by "\?", and any original "\" must be replaced by "\\". The standard does not explain whether this rule, and the corresponding rule introduced for the sake of AnyOrNone, applies only in the constraints part (which is the authors' guess) or also to character strings used in the other parts of the standard, e.g. to initial values of variables in the declarations part or in assignments in the dynamic part. The representation of AnyOne leaves the syntax for literal character strings unchanged and modifies their semantics instead.

Within a SEQUENCE_OF or a SET OF value, AnyOne can be confused with AnyValue, but without any difference in the meaning.

AnyOrNone may be used in a literal constraint value of a string, SEQUENCE OF or SET OF type. It is written as an asterisk. Any number (including 0) of characters, sequence elements or set members, respectively, matches this pattern. For example, "AB*DE" is matched if "ABDE", "AB?DE", or

"AB??DE", etc. is matched (? is AnyValue, see above). To prevent ambiguity, any character "*" in a character string must be represented by "*".

AnyOrNone ::= "*"

A MATCHING MECHANISM FOR SETS, STRINGS AND SEQUENCES

Length attributes are a language element which further restricts possible constraint values of types SEQUENCE OF, SET OF and (any) string by stating an exact value or range for the permitted number of elements. According to 9646, it is used as an attribute for either a specific value or a Complement, Omit, AnyValue, AnyOrOmit, AnyOne, AnyOrNone or Permutation of one of the types listed above. It is written after the expression it modifies, and is enclosed by square brackets.

ValueLength :: = *SingleValueLength* | *RangeValueLength*
SingleValueLength ::= "[" *ValueBound* "]"
*RangeValueLength** ::= "[" *ValueBound* (***TO*** | "..") (*ValueBound* | ***INFINITY***) "]"
ValueBound :: = *Number* | *TS_ParIdentifier* | *TS_ConstIdentifier* | *FormalParIdentifier*

We do not know why Length should not be permitted with Superset and Subset, where it could be used to restrict the size of matching sets equally well as with Complement. Moreover, we cannot recommend to use Length in conjunction with Omit (cf. below) or a specific value, because with these there is no unambiguous intuitive meaning, let alone a formal definition in the standard. Does, for example, "Omit [4]" in an optional field of type BITSTRING simply mean "this field must be absent", or rather "any bitstring, but not of length 4"? Or, is it a constraint that cannot be matched at all, as is (probably) the case with " "01010" [4] "?

MATCHING MECHANISMS FOR OPTIONAL COMPONENTS

Before examining the matching mechanisms for optional components, let us recall that the components explicitly or implicitly declared as optional are

- elements of structured types,
- parameters of tabularly defined ASP types,
- fields of tabularly defined PDU types, and
- ASN.1 OPTIONAL and DEFAULT components.

The TTCN standard permits the Omit, AnyOrOmit and IfPresent mechanisms described below for ASP and CM parameters or PDU fields declared as optional. It does not clearly state whether they may also be used for optional components of a structured type or ASN.1 constraint, but there seems to be no reason why they should not.

Omit is written in capital letters in ASN.1 values and as a dash symbol in TTCN tabular values. It is matched if the relevant component is absent. As noted before, ASN.1 also permits the simple omission of an optional value in its notation of specific values.

*Omit** ::= "-" | ***OMIT***

AnyOrOmit is written as an asterisk. It is matched if the relevant component is either present and consistent with the given type, or if it is absent.

AnyOrOmit ::= "*"

Fortunately, in spite of using the same symbol, AnyOrOmit cannot be confused with AnyOrNone, because the latter is syntactically treated like a string character, SEQUENCE_OF element or SET_OF member. Such components, though in a broader sense optional, are not actually *declared* as optional, as described above.

The *IfPresent* attribute can follow any other matching mechanism that is either without an attribute, or with only a length attribute. It modifies the constraint value with which it is associated, by also permitting the component to be absent.

RELATIONS BETWEEN MATCHING MECHANISMS

Several equivalent ways of expressing particular constraints with matching mechanisms follow from the theorems of mathematics, such as

SUBSET(SUBSET(X)) ≅ SUBSET(X) and
OMIT IF_PRESENT ≅ OMIT.

Some of these equivalences even permit complete replacement of some matching mechanisms by combinations of others, e.g.

OMIT ≅ COMPLEMENT(?)
? ≅ COMPLEMENT(OMIT)
* ≅ ? IF_PRESENT,

where * is meant as AnyOrOmit and ? as AnyValue.

3.5.3 *Tabular constraint declarations*

In this subsection we deal with purely tabular constraint declarations, i.e. those which do not use ASN.1. They serve to specify constraints for structured as well as tabular ASP, PDU and CM types. We will discuss base constraints, whose specifications are closely related to those for the corresponding type definitions, and modified constraints, for which only the modified components are displayed. We will also take a look at constraint chaining, which deals with the constraint-subconstraint relationship. The topic of multiple constraint declarations, a more compact form of presentation, is treated in Subsection 3.5.5.

ASP CONSTRAINT DECLARATIONS

A tabular ASP constraint declaration (see Table 3.46) assigns a previously defined tabular ASP type to the constraint. The constraint then has the same component structure and substructure as the type specified in its constraint declaration, including macro entries. Although the standard is not perfectly clear about this, it seems that short ASP parameter names introduced in the type definition may be used without adding the standardized parameter names in parentheses.

In a base constraint, the derivation path is empty, and all ASP parameters are specified. In a modified constraint, the derivation path lists the ancestors of this constraint in its modification history, and only the parameters which are changed with respect to the parent constraint are listed. Similar rules hold for the other (PDU etc.) kinds of constraints. We will not mention them in the other cases, but will reopen the topic at the end of this subsection.

TABLE 3.46. ASP constraint declaration

<table>
<tr><th colspan="3">ASP CONSTRAINT DECLARATION</th></tr>
<tr><td colspan="3">CONSTRAINT NAME : ConsId&ParList
ASP TYPE : ASP_Identifier
DERIVATION PATH : [DerivationPath]
a “dotted list” of constraint identifiers: {ConstraintIdentifier "."}+
COMMENTS : [FreeText]</td></tr>
<tr><th>PARAMETER NAME</th><th>PARAMETER VALUE</th><th>COMMENTS</th></tr>
<tr><td>:
ASP_ParIdOrMacro
the same as in the corresponding position in the type definition
:</td><td>:
ConstraintValue&Attributes

:</td><td>:
[FreeText]

:</td></tr>
<tr><td colspan="3">DETAILED COMMENTS: [FreeText]</td></tr>
</table>

Value entries obey the following grammar:

*ConstraintValue&Attributes** ::= (*ConstraintExpression* | *MatchingSymbol* | *Consref*) [*ValueLength*] [***IF_PRESENT***] [*ASN1_Encoding*]

MatchingSymbol ::= *Complement* | *Omit* | *AnyValue* | *AnyOrOmit* | *ValueList* | *ValueRange* | *SuperSet* | *Subset* | *Permutation*

Table 3.47 shows a fictitious example of an ASP base constraint declaration: A C-DATArequest service primitive carries a CR (Connect Request) PDU of the next higher layer, called D. This is also an example of static chaining.

TABLE 3.47. An ASP constraint declaration example

ASP CONSTRAINT DECLARATION		
CONSTRAINT NAME : C_DATAreq_C1 ASP TYPE : C_DATAreq DERIVATION PATH : COMMENTS :		
PARAMETER NAME	PARAMETER VALUE	COMMENTS
Calling_Add	TS_Par_From	1)
Called_Add	TS_Par_To	2)
Data	D_CR_1	3)
DETAILED COMMENTS: (very detailed) 1) The ASP parameter Calling_Add obtains its value from the TS parameter TS_Par_From. 2) The ASP parameter Called_Add obtains its value from the TS parameter TS_Par_To. 3) The ASP parameter Data obtains its value from the CR DPDU constraint D_CR_1.		

PDU CONSTRAINT DECLARATIONS

A tabular PDU constraint declaration assigns a tabular PDU type to the constraint and gives constraint values to the PDU fields, in the form shown in Table 3.48. The header lines and the column about encoding are optional; if they are omitted, the default encoding applies.

The encoding-related BNF rules are:

EncVariationCall ::= *EncVariationIdentifier* [*ActualParList*]

*PDU_FieldEncodingCall** ::= *EncVariationCall* | *InvalidFieldEncodingIdentifier* (*ActualParList* | "( )")

Conventions similar to those just described for ASP constraints apply to PDU constraints. In addition, for each PDU type defined in a test suite there must be at least one constraint.

TABLE 3.48. PDU constraint declaration

PDU CONSTRAINT DECLARATION			
CONSTRAINT NAME : *ConsId&ParList* PDU TYPE : *PDU_Identifier* DERIVATION PATH : [*DerivationPath*] ENCODING RULE NAME : [*EncodingRuleIdentifier*] ENCODING VARIATION : [*EncVariationCall*] COMMENTS : [*FreeText*]			
FIELD NAME	FIELD VALUE	FIELD ENCODING	COMMENTS
: *PDU_FieldIdOrMacro* :	: *ConstraintValue& Attributes* :	[*PDU_FieldEncoding Call*]	: [*FreeText*] :
DETAILED COMMENTS: [*FreeText*]			

As a PDU constraint example, the fictitious "DPDU" referenced and chained to a C-ASP constraint in Table 3.47 might be declared as shown in Table 3.49.

CM CONSTRAINT DECLARATIONS

Tabular CM constraint declarations (see Table 3.50) resemble tabular ASP constraint declarations. They differ from the latter in same way as CM type definitions differ from ASP type definitions.

TABLE 3.49. A PDU constraint declaration example

PDU CONSTRAINT DECLARATION		
CONSTRAINT NAME : D_CR_1 PDU TYPE : CR_DPDU DERIVATION PATH : COMMENTS :		
FIELD NAME	FIELD VALUE	COMMENTS
Source_Ref	TS_Par_Source	1)
Dest_Ref	TS_Par_Dest	2)
Variant	A	3)
DETAILED COMMENTS: (very detailed) 1) The PDU field Source_Ref obtains its value from the TS parameter TS_Par_Source. 2) The PDU field Dest_Ref obtains its value from the TS parameter TS_Par_Dest. 3) The PDU field Variant is selected as A, denoting the "protocol variant."		

TABLE 3.50. CM constraint declaration

CM CONSTRAINT DECLARATION		
CONSTRAINT NAME : *ConsId&ParList* CM TYPE : *CM_Identifier* DERIVATION PATH : [*DerivationPath*] COMMENTS : [*FreeText*]		
PARAMETER NAME	PARAMETER VALUE	COMMENTS
: *CM_ParIdentifier* the same as in the corresponding position in the type def. :	: *ConstraintValue&Attributes* : 	: [*FreeText*] :
DETAILED COMMENTS: [*FreeText*]		

STRUCTURED TYPE CONSTRAINT DECLARATIONS

An analogous form is used to declare structured type constraint declarations, cf. Table 3.51. Encoding rules are not specified for structured types, only variations or invalid field encodings may be provided.

TABLE 3.51. Structured type constraint declarations

STRUCTURED TYPE CONSTRAINT DECLARATION		
CONSTRAINT NAME : *ConsId&ParList* PDU TYPE : *StructIdentifier* DERIVATION PATH : [*DerivationPath*] ENCODING VARIATION : [*EncVariationCall*] COMMENTS : [*FreeText*]		
ELEMENT NAME	ELEMENT VALUE	COMMENTS
: *ElemIdentifier* :	: *ConstraintValue&Attributes* :	: [*FreeText*] :
DETAILED COMMENTS: [*FreeText*]		

MODIFIED CONSTRAINTS

In many cases, constraints differ only by a single entry or a few entries. The complete specification of all these similar constraints, including their identical parts, can be avoided either by using a parameterized constraint or by using the *modification* mechanism for constraints.

A base constraint can be modified in an arbitrary number of positions. The resulting modified constraint can again be modified, etc. A modified constraint is declared similarly to its parent constraint, in particular with the same parameter list, except for the following differences:

- It is given a new name.
- The derivation path of the modified constraint consists of the derivation path of the parent constraint (which is empty for base constraints), followed by the name of the parent constraint and a dot. Hence it lists all modification ancestors in descending priority order as a "dotted list".
- Only those components are specified whose values differ from those of the parent constraint. Their values overwrite the respective values of the parent constraint. The other component values are inherited from the parent constraint.

Tables 3.52 through 3.54 show an example of repeated constraint modification.

TABLE 3.52. A simple base constraint declaration

PDU CONSTRAINT DECLARATION		
CONSTRAINT NAME : CONS PDU TYPE : PDU_A DERIVATION PATH : COMMENTS : a base constraint		
FIELD NAME	FIELD VALUE	COMMENTS
Field_1	100	declared as, say, INTEGER in the PDU type declaration of PDU_A
Field_2	'00000000'B	declared as BITSTRING
Field_3	TRUE	declared as BOOLEAN
Field_4	"OK"	declared as PrintableString

Although, strictly speaking, the requirement to list ancestors beyond the parent constraint enforces redundancy, it is conceivable that the lookup of inherited values, as well as compile-time checking of the validity of the derivation path, may be facilitated.

New component values can be any constraint values of the proper type, in particular matching mechanisms. It should be remembered that a modified con-

TABLE 3.53. A modified constraint declaration

<table>
<tr><th colspan="3">PDU CONSTRAINT DECLARATION</th></tr>
<tr><td colspan="3">CONSTRAINT NAME : CONS_MODIF
PDU TYPE : PDU_A
DERIVATION PATH : CONS.
COMMENTS : a modification of CONS</td></tr>
<tr><th>FIELD NAME</th><th>FIELD VALUE</th><th>COMMENTS</th></tr>
<tr><td>Field_1</td><td>200</td><td>This value replaces the old value of 100.</td></tr>
</table>

TABLE 3.54. A re-modified constraint declaration

<table>
<tr><th colspan="3">PDU CONSTRAINT DECLARATION</th></tr>
<tr><td colspan="3">CONSTRAINT NAME : CONS_MODMODIF
PDU TYPE : PDU_A
DERIVATION PATH : CONS.CONS_MODIF.
COMMENTS : a modification of CONS_MODIF</td></tr>
<tr><th>FIELD NAME</th><th>FIELD VALUE</th><th>COMMENTS</th></tr>
<tr><td>Field_2</td><td>–</td><td>This field is to be omitted.</td></tr>
<tr><td>Field_3</td><td>FALSE</td><td>This value overwrites the old value of TRUE.</td></tr>
<tr><td colspan="3">DETAILED COMMENTS:
Field_1 evaluates to 200, as in the parent constraint CONS_MODIF.
Field_4 evaluates to ’"OK", as implicitly in CONS_MODIF and explicitly in CONS.</td></tr>
</table>

straint asking for the omission of a value that is present in its parent constraint will do so by stating the new "value" Omit (written as "-") after the component name. Omitting the entire component leads, by inheritance, to the maintenance of the given value.

Extensive use of repeated constraint modifications, as opposed to last minute assignments, will force the human reader who is trying to find the actual constraint values to numerous lookups in the declarations part.

3.5.4 ASN.1 constraint declarations and ASN.1 oriented references

Constraints for ASN.1 defined types are also specified by means of ASN.1. Like tabular constraints, ASN.1 constraints are either base constraints or modified constraints, and in the latter case the unmodified, inherited components are not listed. The presentation of modifications is adapted to the form of ASN.1 values, using the *replacement* language element. Constraint chaining is handled exactly as in the tabular case. Multiple ASN.1 constraint declarations are treated in Subsection 3.5.5.

As in the case of tabular constraints, TTCN distinguishes between ASP, PDU, CM and "type" constraints. As their tables are quite similar, we show their general pattern in a single table, Table 3.55. Each set of alternative table lines applicable to ASPs, PDUs, CMs, or types, respectively, is shown in angle brackets, separated by "or."

TABLE 3.55. ASN.1 (ASP, PDU, CM, Type) constraint

⟨ ASN.1 ASP CONSTRAINT DECLARATION or ASN.1 PDU CONSTRAINT DECLARATION or ASN.1 CM CONSTRAINT DECLARATION or ASN.1 TYPE CONSTRAINT DECLARATION ⟩			
CONSTRAINT NAME	:	*ConsId&ParList*	
⟨ ASP TYPE	:	*ASP_Identifier*	or
PDU TYPE	:	*PDU_Identifier*	or
CM TYPE	:	*CM_Identifier*	or
ASN.1 TYPE	:	*ASN1_TypeIdentifier*	⟩
ENCODING RULE NAME	:	[*EncodingRuleIdentifier*]	– This line only with PDUs
ENCODING VARIATION	:	[*EncVariationCall*]	– This line only with PDUs or Types
DERIVATION PATH	:	[*DerivationPath*]	
COMMENTS	:	[*FreeText*]	
CONSTRAINT VALUE			
ConstraintValue&Attributes | *Replacement* {"," *Replacement*}			
DETAILED COMMENTS: [*FreeText*]			

The constraint value is based, directly or via modifications, on an ASN.1 value. Attributes can restrict the number of elements in a string, sequence or set (*ValueRange)*, or permit the absence of the specified OPTIONAL or DEFAULT value (***IF_PRESENT***), or specify an encoding variation or invalid encoding for a component. The latter two specifications are accomplished in ASN.1 constraints by including an *ASN1_Encoding* directive following the usual ASN.1 constraint value, described by the following BNF rules:

ASN1_Encoding ::= ***ENC*** *PDU_FieldEncodingCall*

*PDU_FieldEncodingCall** ::= *EncVariationCall* | *InvalidFieldEncodingIdentifier* (*ActualParList* | "( )")

Modification is achieved by stating a *Replacement* instead of a *ConstraintValue&Attributes* as the constraint value:

Replacement ::= (***REPLACE*** *ReferenceList* ***BY*** *ConstraintValue&Attributes*) | (***OMIT*** *ReferenceList*)

ConstraintValue&Attributes is shown expanded after Table 3.46. A *ReferenceList* identifies an access path to a component within an ASN.1 constraint according to the BNF rules

*ReferenceList** ::= (*ArrayRef* | *ComponentIdentifier* | *ComponentPosition*) {*RecordRef* | *ArrayRef* | *BitRef*}

ComponentPosition ::= "(" *Number* ")"

*ArrayRef** ::= "**.** [" *Expression* "]"

*BitRef** ::= "**.**" (*BitIdentifier* | "[" *Expression* "]")

*RecordRef** ::= "**.**" (*ComponentIdentifier* | *ComponentPosition* | *PDU_Identifier* | *StructIdentifier*)

Thus, such a reference list may contain:

- *ArrayRef*, i.e. a dot followed by a component number of a SEQUENCE OF or SET OF component, enclosed by parentheses,
- *ComponentIdentifier*, i.e. a field selector identifier of the corresponding SEQUENCE, SET or CHOICE type definition,
- *ComponentPosition*, i.e. a component number of a SEQUENCE, SET or CHOICE component without identifier, enclosed by square brackets,
- *BitRef*, i.e. a dot followed either by a bit string identifier or by a bit string position enclosed in square brackets,
- *PDU_Identifier* and *StructIdentifier*,

such that all identifiers and parenthesized numbers are separated by dots. Note that a leading *ArrayRef* makes the reference list start with a dot. Component or bit numbers are counted from 0 (for the first component or the leftmost bit). The access path must, of course, exist in the constraint to which it refers.

The omission of an OPTIONAL or DEFAULT component in a base constraint can be expressed by actually omitting it. The omission of a component in a modified constraint, which is present in the parent constraint, is expressed by OMIT <access path of the component>.

Let us illustrate these referencing and modification facilities with a few examples:

Assume an ASN.1 base constraint is declared according to the type (in TTCN, hence with underscores):

```
SEQUENCE{   field_1  INTEGER,
            field_2  BITSTRING,
                     OCTET STRING OPTIONAL,
            field_4  SEQUENCE OF INTEGER    }
```

Assume further, another constraint is to be defined as a modification of this base constraint. Then:

- The "first" bit of field_2 of the base constraint may be redefined by writing "REPLACE field_2 . [0] BY 0", using an access path with *Component-Identifier* and numerical *BitRef.*
- The third component of the base constraint, i.e. the "unidentified" octet string, may be omitted by writing "OMIT (2)", using an access path with *ComponentPosition.*
- the second sequence element of field_4 may be redefined by writing "REPLACE field_4 . [1] BY 5", using an access path with *Component-Identifier* and *ArrayRef.*

Particular bytes of the optional octet string *cannot* be referenced by an ordinary access path, but one can of course define new operations for this purpose.

3.5.5 *Compact constraints declarations tables*

Up to this point, each constraint declaration required a separate table. TTCN also allows tables in which several constraint declarations are grouped together in a compact form. This allows not only space savings, but also an easier overview of closely related constraints, e.g. in the case of modifications. The compact tables are merely graphical variants of the previously described long forms. There are no compact table versions in TTCN.MP.

If the number of constraints and constraint components permits it, and if there are no particular comments related to single components, then several ASP constraint declarations can be represented in a single table. The same holds true for a set of PDU constraint declarations, a set of CM constraint declarations, and a set of structured type constraint declarations. Using the same approach employed for the table pattern for ASN.1 constraints in the previous subsection, we present the table patterns in a single format in Table 3.56. It represents a compact table for *m* constraints with *n* components each. Once again, the head-

TABLE 3.56. Compact tabular constraints declarations

<table>
<tr><td colspan="6">⟨ ASP CONSTRAINTS DECLARATIONS <u>or</u>
PDU CONSTRAINTS DECLARATIONS <u>or</u>
CM CONSTRAINTS DECLARATIONS <u>or</u>
STRUCTURED TYPE CONSTRAINTS DECLARATIONS ⟩</td></tr>
<tr><td colspan="6">⟨ ASP TYPE : *ASP_Identifier* <u>or</u>
PDU TYPE : *PDU_Identifier* <u>or</u>
CM TYPE : *CM_Identifier* <u>or</u>
STRUCTURED TYPE : *StructIdentifier* ⟩</td></tr>
<tr><td rowspan="2">CONSTRAINT NAME</td><td rowspan="2">DERIVATION PATH</td><td colspan="3">⟨ PARAMETER NAME <u>or</u>
FIELD NAME <u>or</u>
PARAMETER NAME <u>or</u>
ELEMENT NAME ⟩</td><td rowspan="2">COMMENTS</td></tr>
<tr><td>*ASP_Par-Identifier*$_1$ <u>or</u>
PDU_Field-Identifier$_1$ <u>or</u>
CM_Par-Identifier$_1$ <u>or</u>
ElemIdentifier$_1$</td><td>...</td><td>*ASP_Par-Identifier*$_n$ <u>or</u>
PDU_Field-Identifier$_n$ <u>or</u>
CM_Par-Identifier$_n$ <u>or</u>
ElemIdentifier$_n$</td></tr>
<tr><td>*ConsId&Par-List*$_1$</td><td>*Derivation-Path*$_1$</td><td>*Constraint-Value&-Attributes*</td><td>...</td><td>*Constraint-Value&-Attributes*</td><td>[*FreeText*]</td></tr>
<tr><td>*ConsId&Par-List*$_2$</td><td>*Derivation-Path*$_2$</td><td>*Constraint-Value&-Attributes*</td><td>...</td><td>*Constraint-Value&-Attributes*</td><td>[*FreeText*]</td></tr>
<tr><td>:</td><td>:</td><td>:</td><td>:</td><td>:</td><td>:</td></tr>
<tr><td>*ConsId&Par-List*$_m$</td><td>*Derivation-Path*$_m$</td><td>*Constraint-Value&-Attributes*</td><td>...</td><td>*Constraint-Value&-Attributes*</td><td>[*FreeText*]</td></tr>
</table>

ers of tables, column groups, and columns have to be picked consistently. Note that no encoding information is provided in compact "tabular style tables."

Table 3.57 provides an example of compact constraints decarations that is practically equivalent to the PDU constraint examples in Tables 3.52 through 3.54. Here we can also see the representation of constraint modifications in compact tables, which is equivalent to the long form.

TABLE 3.57. A compact constraints declarations example

PDU CONSTRAINTS DECLARATIONS						
PDU TYPE : PDU_A						
CONSTRAINT NAME	DERIVATION PATH	FIELD NAME				COMMENTS
		Field_1	Field_2	Field_3	Field_4	
CONS		100	'0000 0000'B	TRUE	"OK"	a base constraint
CONS_ MODIF	CONS.	200				a modification of CONS
CONS_ MODMODIF	CONS. CONS_MODIF.		–	FALSE		a modific. of CONS_MODIF

The tables for compact ASN.1 constraints declarations follow the pattern shown in Table 3.58. Encoding information may be contained in the ASN.1 values in the form of ASN.1 encodings.

TABLE 3.58. Compact ASN.1 constraints declarations

⟨ ASN.1 ASP CONSTRAINTS DECLARATIONS or ASN.1 PDU CONSTRAINTS DECLARATIONS or ASN.1 CM CONSTRAINTS DECLARATIONS or ASN.1 TYPE CONSTRAINTS DECLARATIONS ⟩	
⟨ ASP TYPE : *ASP_Identifier* or PDU TYPE : *PDU_Identifier* or CM TYPE : *CM_Identifier* or STRUCTURED TYPE : *StructIdentifier* ⟩	
CONSTRAINT NAME	ASN.1 VALUE
ConsId&ParList$_1$	*ConstraintValue&Attributes*$_1$
:	:
ConsId&ParList$_n$	*ConstraintValue&Attributes*$_n$

Table 3.59 shows an ASN.1 example corresponding to Table 3.57.

TABLE 3.59. A compact ASN.1 constraints declarations example

ASN.1 PDU CONSTRAINTS DECLARATIONS	
PDU Type : PDU_A	
CONSTRAINT NAME	ASN.1 VALUE
CONS	{ field_1 100, field_2 '00000000'B, field_3 TRUE, field_4 "OK" }
CONS_MODIF	REPLACE field_1 BY 200
CONS_MODMODIF	OMIT field_2, REPLACE field_3 BY FALSE

4.1 Fundamentals of test case behavior description

In this section, we want to introduce the most important concepts of test case behavior description and test case execution. In later sections of this chapter we will supply the details left out in this overview.

In order to avoid monotonous repetitions, we will often refer, not only here, but throughout this chapter, simply to "the tester," instead of "the tester (in non-concurrent TTCN) or a test component (in concurrent TTCN)." Most of what is written about the tester in conventional TTCN holds true – with the due changes, like considering CPs in addition to PCOs – for each test component in concurrent TTCN. Particular roles of test components, for instance with respect to verdict assignment, will be explained in the various contexts in which these roles are treated. We will summarize the main differences in 4.1.11.

Fundamentally, the tester, executing a test case, can be considered as behaving like one sequential process; this is the way its behavior is described in the behavior trees in conventional TTCN. However, on closer examination, this process is made up of a set of cooperating concurrent subprocesses – even in non-concurrent TTCN. One of these, let us call it the *core* process, executes the behavior tree and communicates with the tester's environment at PCOs. The tester core cooperates with the other tester subprocesses, the *timers*. It performs timer operations on them, the timers post timeout notifications in a list, and the core searches for particular timeout notifications in this list and removes them. More details of the tester process structure will be developed in the following subsections and will be summarized in Figure 4.2.

4.1.1 Behavior trees

The central concept of conventional test case specification is the *behavior tree*. In speaking of *the* behavior tree, we abstract – at least for the moment – from the fact that this tree is usually divided into several subtrees, which we will call TTCN trees. The multitude of TTCN trees contributing to one behavior tree in conventional TTCN does not imply any distribution or concurrency; it is nothing more than a different representation of a single tree.

In concurrent TTCN, there is one behavior tree being executed in each active test component. All these behavior tree executions follow almost the same rules, with a few differences between the main test component and the parallel test components that we will explain later.

The elements of a behavior tree are *behavior lines*. In the majority of cases, a behavior line describes an *event,* i.e. the reception or sending of a data object by the tester. The behavior tree of a test case describes event sequences by sharing their common initial segments. For example, the set of the three maximal event sequences shown in the top block of Figure 4.1 can be described by the behavior tree depicted in the middle of this figure. In TTCN, such a tree is not represented by such graphical elements, but rather by means of *indentation*, as shown on the bottom of Figure 4.1. In this notation, lists of alternatives are written left-aligned, one below – but not necessarily immediately below – the other. Sequential events are notated by increasing indentation, the alternative subsequent events equally indented below their common predecessor.

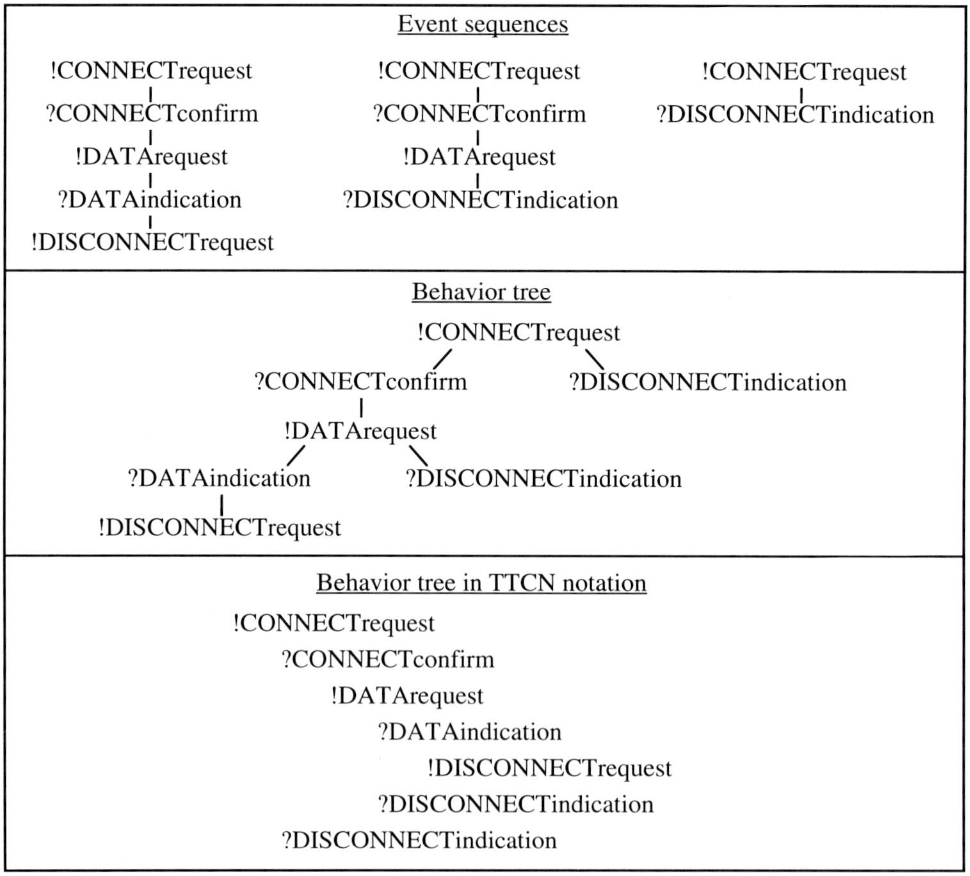

FIGURE 4.1. Event sequences, behavior tree, and TTCN notation

Let us point out here that, strictly speaking, a behavior tree often is not exactly a tree with behavior lines as its nodes. The reason for this is that it may have more than one alternative at the top, while a tree has only a single root node. There are several ways to reconcile practical usage with formal rigor in this respect, e.g.,

- to view the behavior tree as a forest (a set of trees), or,
- to introduce a title for the tree and to consider this title as the root node, or,
- to consider the behavior lines not as the nodes, but as the edges of the tree; in which case the tree's nodes might be identified with states of the test case, and these literally have to be read "between the lines."

It is left to the readers to decide which of these formal views they prefer. We will continue to speak simply of "trees." However, we prefer the term *list of alternatives* to the 9646 term *set of alternatives*, because the order of alternatives is significant: they need not mutually exclude each other, and the operational rule that they are attempted in their given order generally will have an impact on the determination of the eventual choice between them.

4.1.2 Events

Protocol specifications deal with *send*, *receive*, and *timeout* events. Timeout is a technical term meaning timer expiration. The same kinds of events are of interest from the tester's viewpoint. Additionally, *otherwise* events are used as a variant of receive events, and *done* events can be observed in concurrent TTCN, cf. 4.1.11. TTCN provides the following kinds of behavior lines to describe these types of events (*event lines*):

- Send lines, marked by a leading exclamation point, as in
  !DATArequest
- Receive lines, marked by a leading question mark, as in
  ?DATAindication
- Otherwise lines, also starting with "?", as in
  ?OTHERWISE
- Timeout lines, also starting with "?", as in
  ?TIMEOUT Clock_1
- Done event lines, also starting with "?", as in
  ?DONE(PT1,PT2)

As another syntactical detail, send, receive and otherwise lines usually start with the name of the PCO or CP at which these events can happen written in front of the "!" or "?", as, for example, in

LT?DATAindication

where the PCO name is "LT." PCO and CP identifiers can be omitted if only one PCO and no CP are used in the chosen test method.

We have not mentioned the representation of concrete data in test cases yet. In fact, it is insufficient to specify merely the type of PDU, ASP or CM in a send or receive line. In send events, all components (i.e. fields and parameters) have to be given specific values, because the tester is expected to behave deterministically. On the other hand, the IUT or the underlying provider may have a certain latitude in its reactions, with respect to both the type and contents of elements produced. Hence, there will be alternatives for the various types of elements that the tester must be prepared to receive next. In receive events, sets of allowed value combinations, i.e. patterns for values, will be described instead of single values. Both purposes, the definition of specific values for send events and of value patterns for receive events, are served by *constraints*. As shown in Section 3.5, constraints are declared in the constraints part, and as shown in more detail later in this chapter, they are referenced in the behavior line by an entry in the "Cref" column consisting of their name and, where applicable, parameters.

In this simplified overview, however, we will often disregard constraints and pretend that the ASPs, PDUs or CMs to be sent or received are already described sufficiently well by their type names.

With these reservations, we will now provide a crude picture of the execution of the behavior tree of a test case. During test case execution, or, in concurrent TTCN, test component execution, a path through a behavior tree is traveled, from the root to a leaf node. This path is determined by the interplay between the behavior tree and the actions and reactions of the tester's environment, which consists principally of the underlying service provider and, most importantly, the IUT (plus, possibly, the other test components). Starting with the top list of alternatives, the tester tries each alternative on the current list of alternatives, if necessary repeatedly, until one of them succeeds. The "successful" alternative is executed, and its "child behavior lines," i.e. the potential sequential successors, become the next list of alternatives that the tester will try. If this list is empty, i.e. the last behavior line executed was a leaf node of the behavior tree, then the test case ends.

In the following subsections, we will describe which kinds of alternatives are permitted in which context, as well as when an alternative fails and when it succeeds. Four kinds of events involve the reception of information, each of which may fail as an alternative if the specified information has not arrived yet: receive, otherwise, timeout and done. Receive and otherwise lines deal with incoming data "sent" by the IUT, the provider or other test components. Timeout lines deal with information on the expiration of timers. Done events deal with information on the termination of parallel test components. If such an event is the only element on an alternative list, then the attempt to match this event line amounts to actually waiting for the information to arrive. Receive, otherwise, and timeout will be the topics of the next two subsections. The done event will be treated in the context of concurrent TTCN (4.1.11). Other subsections will be devoted to send events and other TTCN lines.

4.1.3 Operational semantics of receive and otherwise events

Let us assume for the moment that the list of alternatives currently considered for the next event comprises only receive lines. At each PCO or CP, the tester can "see" the first element of the input queue, i.e. the "oldest" element put into the queue and not yet removed. For each input queue, the tester takes a note of its first element, or, by substitution, of the fact that this queue is empty. This view of the state of the environment represents a kind of snapshot, which prompted the term *snapshot semantics.* The tester then runs through the list of alternatives, trying to find the first alternative matching the snapshot, i.e. specifying the receipt of a data object at a PCO or a CP that is actually the first element of the associated input queue visible in the snapshot. Matching requires that the data object is of the type whose name is given in the receive line, and that its value matches the pattern defined by the constraint referenced in that line.

The first matching alternative succeeds. The tester removes the matched element from its queue, and moves one level down the behavior tree, after the successful alternative, ready to take another snapshot and to determine the next matching alternative.

If none of the alternatives on the current list matches, it may be caused by the input queue(s) of interest being empty in the snapshot, i.e. none of the expected elements has arrived yet. In that case, after the unsuccessful pass through the list of alternatives, the tester takes a new snapshot, possibly including more events that have arrived, and tries to match the alternatives against this new snapshot. In principle, this cycle is repeated until an alternative matches the last snapshot taken. Obviously, if none of the precautions described in the following text are taken, this cycle can go on indefinitely, specifically if none of the elements expected ever becomes visible at the PCOs and CPs considered in the current list of alternatives.

If all the input queues considered do have elements, but these cannot be matched in one pass through the alternatives, then the impossibility of success with the given alternatives can be detected in finite time. Such a situation would constitute a test case error, because the test case has to be able to handle undesired inputs.

If, however, some of the queues considered simply remain empty forever and nothing matches on the other queues, then this constitutes a test case error that cannot be perceived in finite time. Endless waiting can and should be avoided by the proper use of timers, as shown later on. Actually, endless waiting need not always be blamed on the test suite writer, as can be seen in the following case: 9646 assumes the correct working of the underlying service provider, which may in real testing not always be fully guaranteed. If the tester is waiting for a reaction of the provider that is required by the provider's specification but is, on a bad day, not produced, then the tester will wait forever. It might be useful not to rely on the provider in this respect but to use timers and to be prepared for invalid input.

TTCN offers language elements to avoid such undesired "deadlocks." In order to enable the tester to match any kind of input, including an infinite variety of invalid inputs, TTCN provides the *otherwise* event. A behavior line with ?OTHERWISE matches every incoming element at the specified PCO, provided this element is included in the current snapshot. In this sense, an otherwise is the most encompassing data pattern. During execution of the otherwise, the matched element is removed from its queue.

Infinite waiting at empty input queues can be avoided by the use of *timeout* events. If the expiration of a timer started earlier is checked within a list of alternatives by "?TIMEOUT Timer_name," then at least this timeout alternative will match in finite time, and the list cannot fail matching attempts indefinitely.

Considering the above, any list of receive event alternatives should ordinarily end with an otherwise event and a timeout event, as in the following example:

```
PCO1?PDU_A
PCO1?PDU_B
PCO1?OTHERWISE
?TIMEOUT Clock_1
```

4.1.4 Operational semantics of timeout events

Timers that have expired do not influence test case execution immediately, i.e. they do not effect an interrupt. Instead, the expiration of a running timer is reported by the addition of a *timeout notification* to a set dedicated for these notifications. 9646 speaks of a list, though the order on this list is irrelevant. The current contents of this list are recorded with each snapshot. When attempting a timeout alternative, the tester scans the recorded list for a notification from the given timer. If one is found, the timeout succeeds, and the corresponding notification is removed. If no timer is named, any timeout notification is accepted, and in case of success all notifications are removed.

Apart from preventing infinite waiting, timeouts can also be used to check whether or not the IUT violates specified time limits. Some experts consider such investigations to be a part of performance testing and hence beyond the scope of conformance testing. In their view, timeouts can only be used to set upper limits dictated by the "tester's patience," regardless of protocol timing requirements. The authors tend, however, to the former view, i.e. that protocol-specific time related behavior is also subject to conformance testing.

The IXIT is the proper place for the duration values of OSI specification independent timers that are agreed upon both by the client and by the test laboratory.

If timeouts are used to check proper timing behavior, some precautions must be taken not to be unreasonably strict. For instance, if a response from the IUT is expected before a particular timeout, and the next snapshot shows both the arrival of the response and the expiration of the timer, then it cannot be determined which of the two events happened first. Note that TTCN semantics do not provide timestamps on queued events. The list of alternatives should therefore contain the reception of the response before the timeout, such that the former is successful with the assumed snapshot. Other reasons for this order are explained below.

By the way, even though the standard does not require it, it probably makes sense to use timers only "locally." We recommend managing and checking a timer in each test case only within one test component, or, in conventional TTCN, exclusively in connection with the activities at only one of the PCOs. Otherwise, the test tool may have to go to great lengths to manage global timer handling. Chances are that without this restriction, timing-related information, like snapshots of the timeout notification list or any time readings, will often be quite outdated due to communication delays.

TTCN forbids the use of timeouts to mask or "to help the tester over" unexpected elements in the input queues: timeouts should only be applied if one of the queues considered is empty. It must be impossible to receive at the PCOs considered, after the timeout, anything other than what could be received before the timeout. The recommended use of otherwise and timeout lines as last alternatives to receive lines implies the observation of this rule. Without this precaution, in more complicated cases, it may be hard to determine whether the rule is fully observed.

4.1.5 Operational semantics of send events

Putting aside the critical discussion of PCOs in 4.9, and without considering qualifiers in behavior lines, we can state that TTCN requires that send events always succeed.

A send event should always be the only member of its list of alternatives, for the following reasons:

- On the one hand, due to the certain success of the send event, any alternative behavior line listed after it will never be reached in the execution of the test case. Such lines would be *unreachable lines* and are to be avoided.
- On the other hand, it is not permitted to let a send event follow a receive, otherwise or timeout event in a list of alternatives. The reason for this restriction is that it might then depend on accidental and temporary timing conditions whether the incoming elements have been put into their respective queue or whether they are still "on their way." This was thought to create an undesirable amount of nondeterminism (or non-repeatability) in testing.

In an invalid test case containing the passage

```
!PDU_Ask_for_A
        ?PDU_A
        !PDU_Ask_again,
```

it would depend on indeterminate timing relationships in the IUT, the tester, and the underlying provider whether the tester sends the "reminder" or not.

4.1.6 Pseudo-events

Let us, for our purposes, use the term *pseudo-event* for assignment lists, qualifiers and timer operations. 9646 uses this term for what we will call pseudo-event lines below and does not have any particular term for our pseudo-events.

Besides processing events, a tester is required to do some calculations, to check some conditions, and to operate timers. Calculations are performed in *assignments*, conditions are checked in *qualifiers*, and timers are managed in *timer operations*. A pseudo-event can form a *pseudo-event line* on its own in the behavior tree. For example, if x and y are integer test case variables, then

```
(x:=1)
```

is an assignment line, and

```
[y>1]
```

is a qualifier line.

An assignment line, like a send line, always succeeds. Therefore, the same restrictions about possible alternatives apply.

A qualifier line has the effect of a switch: If it evaluates to TRUE it succeeds, and the behavior lines for subsequent behavior are treated next; otherwise it fails. In the latter case, either the next alternative on the list is attempted, or, in the case where there is none, the entire list of alternatives is tried again with a new snapshot.

Usually, a qualifier will not be the only alternative in the list. It is probably expected to fail sometimes – or it would not be used there in the first place. However, if a single-alternative qualifier fails, the test case gets stuck there, because the qualifier is retried forever with identical values. Single-alternative qualifiers are not generally forbidden in TTCN, as shown by the "dummy line" [TRUE], which is used occasionally in connection with constructs, cf. Section 4.5.

As to timer management, let it suffice for this general overview to state that there exist *timer operations* to start, to cancel and to read declared timers. Details on timer operations, which are in fact operations executed by the tester on timers, are given in 4.3.4.

Several pseudo-events can be combined to form a pseudo-event line. One or more pseudo-events can also be added to event lines. Multiple assignments are listed within one pair of parentheses, separated by commas. Multiple timer operations are also separated by commas. Multiple qualifiers are not allowed, but their effect can be achieved by using the logical conjunction of the qualifiers as a single qualifier. A valid combination of pseudo-events contains at least one pseudo-event, and, in particular, at most one qualifier and zero or more assignments and/or timer operations, in that order. The effect of a pseudo-event line is shown by the following "behavior expansions," written informally but in TTCN tree style:

A part of the form

```
[qualifier]  (assignment list)  timer operations
    <subsequent behavior if qualifier true>
<next alternative if qualifier false>
```

within a behavior tree could be replaced by

```
[qualifier]
    (assignment list)
        timer operations
            <subsequent behavior if qualifier true>
<next alternative if qualifier false>
```

Assignment lists and lists of timer operations are evaluated from left to right, such that

(assignment$_1$, ..., assignment$_m$)

could be replaced by

(assignment$_1$)

...

(assignment$_m$)

and likewise

tim_op$_1$, ..., tim_op$_n$

could be replaced by

tim_op$_1$

...

tim_op$_n$.

These details matter in regard to the overall effect of a combination of pseudo-events. For example, the precedence of assignments before timer operations matters, because reading a timer includes storing the obtained value in a variable which may also appear on the right-hand side of an assignment.

We will explain the effect of pseudo-events adjoined to event-lines later, because it involves a few subtle details, see Subsection 4.4.4. For now we merely note that

- qualifiers on event lines are considered "ahead of anything else," i.e. if they fail, the line fails, and if they succeed, they can be ignored, and the line is treated as if they were not there;
- assignments and timer operations on event lines are only performed if the line succeeds.

This will suffice to understand the example in Subsection 4.1.8. In principle, TTCN test cases could be expressed without mixing events and pseudo-events on one line [BAER 94], but this would increase the number of lines.

4.1.7 The tester's abstract architecture and abstract behavior

With the descriptions of send, receive and timeout events and of assignments and qualifiers given up to this point, we can construct the rough architectural sketch of a "tester process," shown in Figure 4.2. The arrows represent atomic interactions of the driving core component on the left with subsidiary processes (timers) and "data repositories" (parameters, constants, variables, PCO and CP queues, timeout notification list). The direction of the arrows informally indicates information flow. A tester process is meant to encompass, in conventional TTCN (i.e. in the single-party context), an LT and possibly a UT; in that case the CPs in the figure should be disregarded. In concurrent TTCN (i.e. in the multi-party context), a tester process corresponds to a test component.

It should be borne in mind that the operational description of the behavior of a test case only applies on an abstract level. The details of operational semantics are by no means binding requirements on the realization of the test case. The standard only requires that the tester executing a test case behave the same way *externally* as if it were executing the test case according to operational semantics. No specific implementational details are implied by that. The tester can communicate with the IUT, with the underlying provider(s) and among its components by any discipline desired, and it can work internally in any manner, as long as it achieves the same effect externally as the operational semantics would.

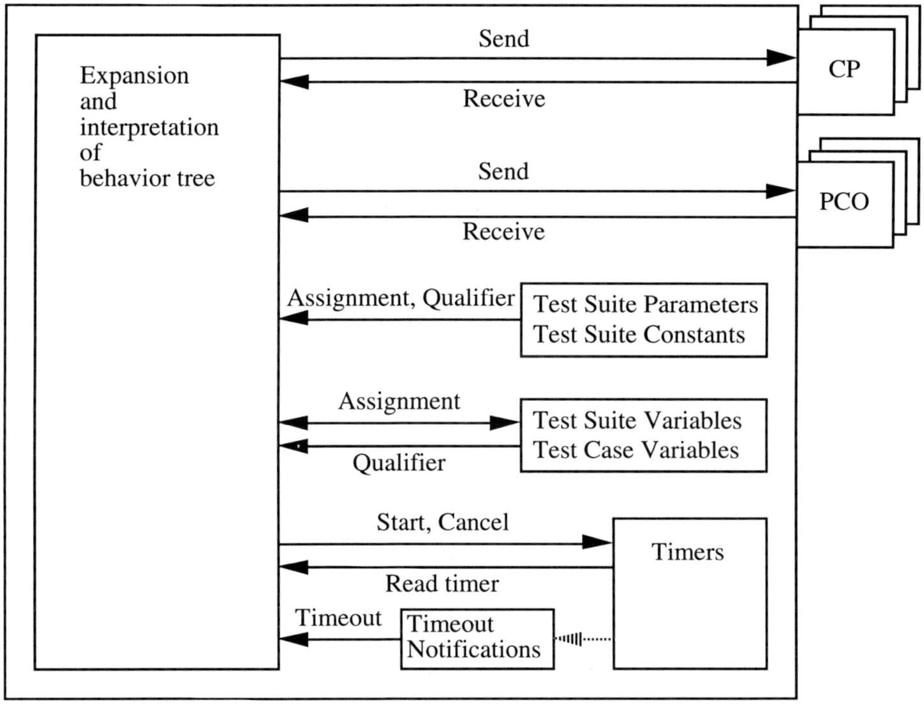

FIGURE 4.2. A simplified conceptual architecture of a tester process

4.1.8 An example of test case execution

We will now illustrate this rough description of the semantics of behavior trees by means of a small example, shown in Figure 4.3. References to constraints are not considered. We deal with only one PCO and no CP and hence with only one input queue.

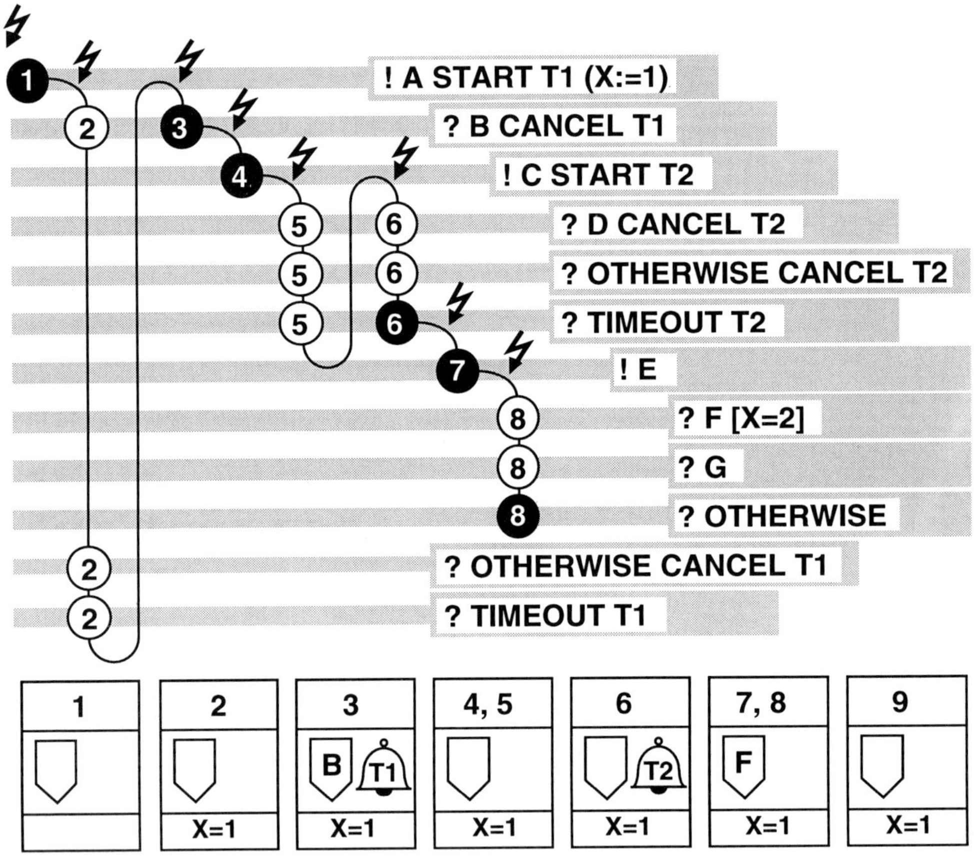

FIGURE 4.3. A tester performing a behavior tree

The lines of the behavior tree are traversed, i.e. attempted or executed, in the order indicated by the solid curved line. Flash symbols represent the taking of a snapshot of the "tester environment," i.e. the input queue and the timeout notification list. With this snapshot, the tester goes, possibly repeatedly, through the next list of alternatives, trying to find a successful behavior line. We depict the contents of the snapshots by the numbered views of the tester environment shown at the bottom of the picture. A bell symbolizes a timeout notification. The arrow bar shows either the first element of the input queue, or nothing if the queue is empty. We consider the contents of parameters, constants and variables as information available to the tester, but not as a part of the snapshot, and we show it attached to the bottom of the snapshot.

Numbers in circles indicate the current snapshot, on the basis of which the tester is currently trying the alternatives. Black circles represent successful behavior lines, for example, a matching receive line.

Note that the behavior tree is neither written in full TTCN, nor is it related to real protocols; it exists only as a compact example for several kinds of behavior lines. Still, we recommend that readers go through this example step by step, to underpin their understanding of the semantics of behavior trees. We describe these steps below. Some of the details about the execution of behavior lines will be explained at greater length later in this chapter.

In the initial snapshot (1) of the tester's environment, there is not any PDU or ASP to be received, nor does there exist any notification that a timer has expired. By convention, the sending of A as well as the associated assignment succeed unconditionally.

Data objects sent off are no longer visible, and the next snapshot (2) shows the same picture. As there is nothing to be received, nor any timeout to be observed, all three alternatives on the next list of alternatives fail.

Snapshot (3) shows that in the interim a B has arrived and that timer T1 has expired. Now, the first alternative succeeds. Note that the other two alternatives would have succeeded, too, but ?B "won" against its "competitors" simply because it was listed first. The execution of ?B includes the removal of B from the input queue, and the execution of CANCEL T1 removes the timeout notification of T1.

In the absence of any other arrivals, snapshot (4) shows what is left after the previous behavior line: nothing. But nothing is needed for the next possible single alternative to succeed, because it is a send line: !C. As a part of the execution of this line, timer T2 is started.

As the next snapshot (5) shows the same state of affairs, all three alternatives on the next list fail.

Snapshot (6) shows a timeout of T2 and nothing at the PCO, thus the timeout alternative succeeds.

In the next snapshot, an F has arrived at the PCO. However, that does not matter for !E to succeed.

Nothing has changed in the next snapshot (8). The first alternative fails, because X=1 and hence X≠2. The second fails, because there is no G at the front of the input queue. The OTHERWISE succeeds on F, removing it from the queue, which is finally empty (9).

4.1.9 TTCN constructs, TTCN trees

In spite of the fact that data object value or pattern descriptions are only referenced from behavior trees, and the descriptions themselves are given separately in the constraints part, test cases for real protocols are usually so complex that their representation by a single behavior tree is at least impractical. In fact, they often require infinitely large behavior trees, which cannot be written down at all without further notational conventions.

This problem is solved by modularization: subtrees of the behavior tree are detached; they are each given a name and are described separately. In the behavior tree, the removed subtree is replaced by an instruction to attach the subtree, an *attach statement,* of the form "+ <name_of_subtree>". As in the case of procedures in programming languages, this course of action permits a more efficient description, because identical subtrees appearing in several places of the same tree or in several trees have to be described only once. Figure 4.4 illustrates the principle of attachment. LongTree can be reduced to ShortTree by detaching the instances of SubTree. It can be reconstructed by attaching an instance of SubTree in the place of each corresponding attach statement.

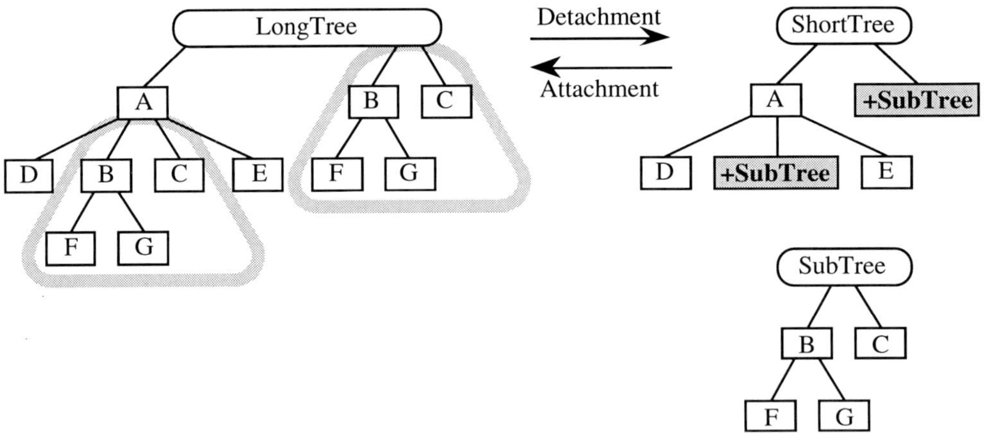

FIGURE 4.4. Modularization of behavior trees

By the recursive use of tree attachment, even an infinite behavior tree can be represented by a finite number of finite trees. The attachment method is even extended by permitting formal parameters in subtrees, such that not only identical subtrees but also subtrees "identical up to parameter values" can be described collectively in one tree.

The attach statement is called a *construct* in TTCN. There are other constructs besides the attach which also may be used to write abridged behavior trees. We will show in Section 4.5 that the goto construct for jumping and the repeat construct for looping can be explained in terms of attachment.

A behavior tree, describing the tester behavior in the execution of a test case, is usually described by one *root tree* of the test case and by several subtrees, referred to in attach statements. We will call all these constituent trees, root tree and subtrees, *TTCN trees*. They contain event lines, pseudo-event lines and constructs.

Subtrees which are specific to a given test case are listed together with the root tree in the test case behavior description; we will call them *test case local trees*. Those which may be attached by several test cases are called *test steps* and are collected in a test step library – at least in simple cases. The true picture is slightly more complicated: test step behavior descriptions are analogous to test cases; they are generally collections of TTCN trees, always comprising one *test step root tree* and possibly additional *test step local trees*. Any TTCN tree may refer to (i.e. attach) any test step, which results initially in the insertion of the test step root tree. We illustrate the various possibilities of TTCN tree attachment in Figure 4.5.

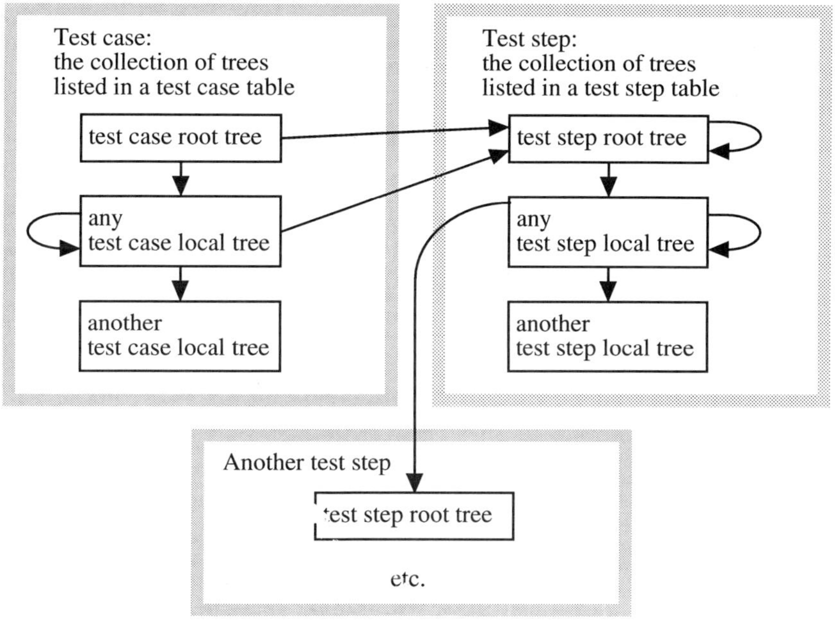

FIGURE 4.5. The possibilities of tree attachment

The four types of TTCN trees have many things in common, but there exist specific differences in syntax and semantics. Wherever these differences do not matter we will try to treat the trees together.

Note that in a few places the standard uses the term "local test steps" for local trees, but without calling the TTCN trees in the test step library "global" test steps. This has to do with the testing methodology's terminology, which speaks only of test steps in general. We will use the TTCN specific terms and differentiate between local trees, which are separate subtrees in test case and test step tables, and test steps, which occupy tables of their own and comprise a test step root tree and possibly test step local trees.

The term *default* is used for a TTCN mechanism which allows attachment of identical subtrees throughout a TTCN tree, without having to use scores of attach statements, let alone to spell out these subtrees everywhere. The default mechanism is used in particular to declare a standard set of additional alternatives to be appended to all lists of alternatives. A typical candidate for this is a uniform "problem treatment," consisting, for example, of an otherwise statement to match unexpected observations, and a timeout statement to match the absence of any expected reaction, both followed by the appropriate behavior to return all participating processes to a stable testing state. Default trees are specified in a default library, and they are referenced in the test case and test step tables. In response to detailed requests from test suite writers, the default mechanism was extended to permit controlled variations in the default trees during test case execution. Details of this feature are presented in Section 4.5.

The operational semantics of TTCN describe not only the tester's path through the behavior tree; they also describe how the tester composes the relevant parts of the behavior tree from the TTCN trees, as directed by the constructs. In the practical execution of tests, the behavior tree will hardly ever be constructed in full detail. Rather, the tester will be operating on a *working tree* that originally consists of the test case root tree and is, step by step, *expanded* only as far as necessary to determine the next list of alternatives. The working tree is thus a dynamic approximation to the behavior tree. Its development is not always the same, because each run of the test case depends on the behavior of the tester's environment. For example, the top right occurrence of +SubTree in Figure 4.4 would always be expanded, in order to determine the top level list of alternatives. The other occurrence of +SubTree between D and E would only be expanded if A succeeds among the first list of alternatives.

Another TTCN construct is used only in concurrent TTCN: the *create* statement. While attach, goto and repeat statements (and defaults) are mechanisms describing how the behavior tree is assembled from the test case's TTCN trees, the create statement initiates the concurrent execution of an additional test component. Each active test component conceptually works on its own behavior tree, i.e. it composes its own working tree. Only the main test component of a test case can create parallel test components.

Now we can summarize the types of lines in the behavior tree and in TTCN trees:

- Event lines (possibly including a valid combination of pseudo-events) may be
  - send lines,
  - receive lines, including otherwise,
  - timeout lines,
  - implicit send lines, or
  - done lines.
- Pseudo-event lines consist of a valid combination (cf. 4.1.6) of pseudo-events. Pseudo-events may be
  - qualifiers,
  - assignments, or
  - timer operations.
- Constructs do not appear in the behavior tree. They appear in TTCN trees and in the working tree, and they may be expanded during the composition of the relevant part of the behavior tree. A construct is either a
  - goto,
  - attach,
  - repeat,
  - activate,
  - return,
  - no-defaults, or
  - create.

Implicit sends are addressed in 4.4.5. Details on constructs and defaults are given in Section 4.5. Defaults, though closely related to the attach, are not behavior lines. They are referenced in behavior table headers.

As this subsection addresses TTCN trees in general, it will be in order to remind of the restrictions on the types of lines and their order in lists of alternatives, addressed in various contexts. Let us assume for the moment that we are dealing with one list of alternatives in which constructs have already been expanded, events and pseudo-events appear on separate lines, pseudo-event lines contain only one pseudo-event each, and lists of qualifier lines are complete, i.e. the qualifiers "add up to" TRUE. All of this can be achieved by transformations preserving the meaning of the test case. Then, if the restrictions referenced above are respected, and garbage lines are avoided, this list is generally of one out of three possible types:

- *Receive type lists* expand (after the removal of constructs) to contain only receive, otherwise, done and timeout lines. Any otherwise line on a PCO will appear only after the last receive on this PCO, and all timeouts appear at the end.
- *Unconditional lists* contain exactly one line which is either an assignment, a timer operation, or a send.
- *Qualifier lists* contain only a number of qualifiers, and the qualifiers preceding any single one will not "add up to" TRUE.

Since TTCN-trees will generally not fulfil the simplifying assumptions made above, lists of alternatives may consist of other combinations, but the above typology shows what a list of alternatives semantically should amount to.

4.1.10 Verdicts

In conformance testing, the ultimate purpose of test case execution is the assignment of a *verdict*. Every test case execution must result in a verdict, which is either *pass*, *fail*, or *inconclusive*. In conventional TTCN this execution corresponds to a path through the behavior tree, and hence the most obvious places to assign verdicts are the leaves of the behavior tree. Alternatively, TTCN permits assignment of preliminary verdicts before the end of the test case, and calculation of the final verdict from these preliminary verdicts. In concurrent TTCN, all parallel test components contribute preliminary verdicts, and the main test component computes and assigns the final verdict.

Both preliminary and final verdicts affect the current value of a predefined *result variable* R, and the final computation merely delivers the latest value of R. In concurrent TTCN, each test component has its own private copy of R, but the MTC's result variable R is globally affected by preliminary results in any test component, and is used to produce the final verdict. More details on test verdicts, in particular their meaning and their assignment, are presented in Section 4.6.

4.1.11 Concurrent TTCN

Only the main test component may use the *create* construct, for example

CREATE(TCompA:StepA(par1,par2), TCompB:StepB),

to start the execution of inactive parallel test components that have been declared and listed in the chosen test configuration. It is only at this moment that a particular behavior description is connected with the test component, the connection being valid only for the given test case. The behavior is assigned to the test component (for example TCompA above) by associating it with a test step or local tree (StepA), including its actual parameters (par1,par2) where necessary.

*Create** ::= ***CREATE*** "(" *CreateTComp* {"," *CreateTComp*} ")"
*CreateTComp** ::= *TCompIdentifier* ":" *TreeReference* [*ActualParList*]

In any test case, each test component is created at most once.
While parameters are passed to attached trees by textual replacement, they are passed to created test components by value.

A test component can determine by means of a *done* event line, such as

?DONE(TCompA,TCompB),

whether each of a specified list of parallel test components have terminated. The MTC may even inquire after the status of all of the PTCs started without explicitly specifying them:

?DONE().

If the response is yes, implying, of course, that the information has become available at the requesting site, then the done line succeeds, and the test component continues with the behavior sequentially following the done line. If the done line fails, then the next alternative on its list is attempted.

At the moment of writing this book, it was not clear how much of the default context (cf. 4.5.5) at the point of creation carries over to the newly started test component. If, for example, a local tree is started by a create, is it endowed with the defaults defined in the behavior table containing this subtree, or with the defaults valid at the time the create statement was executed?

We will use the remainder of this section to summarize the special features of concurrent TTCN, cross-referencing the various passages where details are provided.

The main test component is the one started with the test case root tree. From the MTC, PTCs may be started with various behaviors by means of create statements. Test components may, but need not, run on different physical systems.

Test components and their possible configurations must be declared in the declarations part. A test component has a fixed number of PCOs and of CPs throughout a test suite. It has the same PCOs and the same CPs (at least as far as the latters' names are concerned) in any particular test component configuration, and hence throughout a test case, because each test case commits itself to one configuration. PCOs and CPs may only be used in test components for which they have been declared in the test component configuration table.

Each test component started obtains its own set of freshly initailized copies of all test case variables. If values of a test case variable are needed in other components, they must be transmitted via CMs. Test suite variables are only used by the MTC. Each test components started also obtains its own copies of all timers.

Information can be passed from one test component to the other in the following ways:

- explicitly by sends and receives of CMs at CPs (see 3.1),
- implicitly e.g. by
  - verdicts in PTCs, which are counted as preliminary verdicts and influence the global R used in the MTC (cf. 4.6.2), and
  - done events, which provide termination data.

Other ways are theoretically possible, but only as minor side effects of mechanisms that are not really intended for test component communication.

The MTC is not only the first test component started, but also the last one to terminate, and the one to pronounce the final verdict , cf. 4.6.2.

We close this overview of concurrent TTCN with an example in which we contrast a small non-concurrent test case for the distributed test method (Table 4.1) with the corresponding concurrent test case for two test components in the positions of the upper and lower tester, respectively (Table 4.2). The test cases are shown in a strongly simplified form; syntax and behavior are incomplete: the table header, many columns, and provisions for IUT errors are omitted. Dots have been added to facilitate determining the indentation.

TABLE 4.1. A non-concurrent test case

U!A_CONNECTreq
. L?B_CONNECTind
. . L!B_CONNECTres
. . . L?B_DATAind
. . . . L!B_DATAreq
. . . . . U?A_CONNECTcon
. . . . . . U!A_DISCONNECTreq
. . . . L?B_DISCONind
. . . . . U?A_DISCONNECTind
. . . . . . L?B_DISCONNECTind
. . . L?B_DISCONNECTind
. . . . U?A_DISCONNECTind
. U?A_DISCONNECTind

TABLE 4.2. A corresponding concurrent test case

```
CREATE(U_TESTER,U_TREE)
.   L?B_CONNECTind
.   .   L!B_CONNECTres
.   .   .   L?B_DATAind
.   .   .   .   L!B_DATAreq
.   .   .   .   .   L?B_DISCONind
.   .   .   L?B_DISCONNECTind
.   .   .   .   DONE(U_TESTER)
.   .   .   DONE(U_TESTER)
.   DONE(U_TESTER)

U_TREE
U!A_CONNECTreq
.   U?A_CONNECTcon
.   .   U!A_DISCONNECTreq
.   U?A_DISCONNECTind
```

The idea behind this test case is to establish and release a connection for the higher layer A, both activities being started by the upper tester. A-Connection establishment is achieved by the establishment of a connection in the underlying layer B and the transport of the A-connection request and response via B as ordinary data. Entities in both layers may disconnect independently, but the IUT is supposed not to disconnect actively in this test case. Disconnections are unconfirmed. Disconnection by an A-entity is reported to the other side via a B-disconnection.

4.2 Behavior tables

4.2.1 The global structure of the dynamic part

The dynamic part is the fourth part of an abstract test suite; it describes the behavior of the test cases by means of TTCN trees. It consists of the test case tables, followed by the test step tables, if any, followed by the default trees, if any:

DynamicPart ::= ***$DynamicPart*** *TestCases* [*TestStepLibrary*]
[*DefaultsLibrary*] ***$End_DynamicPart***

All three collections, test cases, test steps and defaults, may, but need not, be endowed with a hierarchical group structure, expressed the same way for each collection. We show this structure in the example of the test cases.

In TTCN.MP, the nesting and the names of the groups are expressed by enclosing each group by an opening ***$TestGroup*** *TestGroupId* and a closing ***$End_TestGroup***. In TTCN.GR test cases, grouping information can only be inferred from the "containing group paths" (*TestGroupReference*) in the headers. Important BNF rules in this context are:

> *TestCases* ::= ***$TestCases*** ({*TestGroup* | *TestCase*}+) ***$End_TestCases***
>
> *TestGroup* ::= ***$TestGroup*** *TestGroupId* {*TestGroup* | *TestCase*}+ ***$End_TestGroup***
>
> *TestGroupReference* ::= [*SuiteIdentifier* "/"] {*TestGroupIdentifier* "/"}

Grouping among the test steps and defaults is achieved in a similar manner, as can be seen from the relevant BNF rules and table patterns.

4.2.2 Test case tables

Table 4.3 shows the pattern for test case tables. As the test case name is required to be unique even without the group reference, the mandatory group information repeats what is already provided in the test case index. The possible configurations, which can be referenced in the test case header, were described in Chapter 3. Default references may be accompanied by a list of actual parameters. Defaults will be treated in detail in 4.5.

The body of the test case table is divided into columns which correspond to the components of the behavior lines, except for the line number column, which has no equivalent in TTCN.MP. In the test case table body, tree headers, written into the Behaviour Description column, may occur between behavior lines, cf. Table 4.3. These headers precede and identify test case local trees. Listing headers in this way avoids interrupting the column structure for each local tree header. The root tree does not start with a header; rather, the test case name given in the table header is considered to be its name.

The BNF for test case table bodies is as follows:

> *BehaviourDescription* ::= ***$BehaviourDescription*** *RootTree* {*LocalTree*} ***$End_BehaviourDescription***
>
> *RootTree* ::= {*BehaviourLine*}+
>
> *LocalTree* ::= *Header* {*BehaviourLine*}+
>
> *BehaviourLine* ::= ***$BehaviourLine*** *LabelId Line Cref VerdictId* [*Comment*] ***$End_BehaviourLine***

TABLE 4.3. Test case table

<table>
<tr><td colspan="6">TEST CASE DYNAMIC BEHAVIOUR</td></tr>
<tr><td colspan="6">TEST CASE NAME : TestCaseIdentifier – must be unique throughout the test suite
GROUP : TestGroupReference – structural information, see Subsection 4.2.1
PURPOSE : FreeText – an informal short description
CONFIGURATION : TCompConfigIdentifier – line omitted completely in non-concurrent TTCN
DEFAULT : [DefaultRefList] – a list of default references, separated by commas cf. Section 4.5
COMMENTS : [FreeText] – this line may perhaps also be omitted completely in the second version of the standard</td></tr>
<tr><td>NR</td><td>LABEL</td><td>BEHAVIOUR DESCRIPTION</td><td>CONSTRAINTS REF</td><td>VERDICT</td><td>COMMENTS</td></tr>
<tr><td>:</td><td>:</td><td>:</td><td>:</td><td>:</td><td>:</td></tr>
<tr><td>m</td><td>[Label]</td><td>StatementLine (indented)</td><td>[ConstraintReference]</td><td>[Verdict]</td><td>[FreeText]</td></tr>
<tr><td>:</td><td>:</td><td>:</td><td>:</td><td>:</td><td>:</td></tr>
<tr><td></td><td></td><td>TreeHeader</td><td></td><td></td><td></td></tr>
<tr><td>n</td><td>[Label]</td><td>StatementLine (indented)</td><td>[ConstraintReference]</td><td>[Verdict]</td><td>[FreeText]</td></tr>
<tr><td>:</td><td>:</td><td>:</td><td>:</td><td>:</td><td>:</td></tr>
<tr><td colspan="6">DETAILED COMMENTS: [FreeText]</td></tr>
</table>

Some column headers may be abbreviated in order to provide more space for the statement lines of the behavior description: "Label" as "L", "Constraints Ref" as "Cref", "Verdict" as "V", and "Comment" as "C".

The *line number* column may be omitted completely, in particular if labels are used extensively. The line numbers count only the statement lines, starting with line number 1. Headers are not counted. Line numbers also do not count the typographical lines resulting from entries spreading over several of these; this is treated in 4.2.6. The use of line numbers is recommended for logging purposes, to facilitate line continuation, and as references in detailed comments. If line numbers are referenced, the absence of the Nr-column in TTCN.MP presents a certain obstacle, but that can be handled by the system (TTCN tool, test system, ATS-ETS transformer, etc.) reading the MP-form.

For reasons of backward compatibility with the DIS TTCN version, the label column may be placed behind the behavior description column, although the "grandfather clause" (cf. Chapter 3) would have sufficed for this purpose. Target lines for goto statements must be referenced by their label, not by their line number which does not appear TTCN.MP. With the proper changes to TTCN, it would be possible to merge line numbers and labels into one concept, sufficient for the purposes of either one.

The behavior description column contains the lines of the TTCN trees with the proper indentation:

Line ::= ***$Line*** *Indentation StatementLine*

Some test suite writers would have preferred vertical alignment of the ?'s of alternative receive and timeout statements, even though it would cause PCO and CP names of different lengths to jut out by various amounts to the left. However, in TTCN.GR indentation refers to the position of the first character of the statement line. Graphical indentation is performed in multiples of an arbitrary indentation unit. In the MP form, the depth of indentation, i.e. the number of these units, is given in non-negative integers: 0, 1, 2, ...:

Indentation ::= "[" *Number* "]"

The references in the "Constraints Ref" column point to ASP, PDU or CM data objects or data object patterns defined in the constraints part of the test suite. We will deal with them in 4.4.1.

Optionally, a verdict may be assigned to the line in the verdict column. Basic concepts of verdicts are explained in 4.1.10; details will be described in 4.6.

The comments column and the detailed comments footer can be omitted completely.

4.2.3 Multiple test case tables

Up to this point, each test case was defined in a separate table. TTCN also allows tables in which several test cases are grouped together in a compact form. These test cases must be one-liners, namely test step attachments, and as such they usually have parameters. They must belong to the same test group, and they must all have the same test case default (or all of them none). Note that various defaults can be assigned in the various test steps called. They must all use the same test component configuration (or none, if they are all in non-concurrent TTCN). Verdicts, labels and, of course, constraint references will only appear within the test steps called. The table pattern is shown in Table 4.4.

Multiple test case tables allow not only space savings, but also an easier overview of closely related test cases. These compact tables are nothing more than graphical variants of the previously described long forms; they have no counterpart in TTCN.MP. Hence, if the grouping of several test cases into one table is to be transferred, then this has to be accomplished by additonal information.

TABLE 4.4. Multiple test case table

<table>
<tr><th colspan="4">TEST CASE DYNAMIC BEHAVIOURS</th></tr>
<tr><td colspan="4">GROUP : TestGroupReference – structural information
CONFIGURATION : TCompConfigIdentifier – line omitted completely in non-concurrent TTCN
DEFAULT : [DefaultRefList] – a list of default references, separated by commas, cf. Section 4.5</td></tr>
<tr><th>TEST CASE NAME</th><th>PURPOSE</th><th>TEST STEP ATTACHMENT</th><th>COMMENTS</th></tr>
<tr><td>:</td><td>:</td><td>:</td><td>:</td></tr>
<tr><td>TestCaseIdentifier</td><td>FreeText</td><td>Attach - cf. 4.1.9</td><td>[FreeText]</td></tr>
<tr><td>:</td><td>:</td><td>:</td><td>:</td></tr>
<tr><td colspan="4">DETAILED COMMENTS: [FreeText]</td></tr>
</table>

4.2.4 Test step tables

A test step table specifies one test step root tree and, possibly, additional test step local trees. Test steps can be referenced in attach and create statements in any TTCN tree within a test case or test step table, though in the case of the create only if the tree is executed by the MTC. In contrast with test cases, test steps may be parameterized. If they are, their name is followed by a formal parameter list in the defining test step table, and by a list of actual parameters wherever the test step is referenced in an attach or create statement:

TestStepId&ParList ::= *TestStepIdentifier* [*FormalParList*]

Table 4.5 shows the pattern for test step tables. In contrast to the test purposes associated with test cases, the aims of test steps and test step groups are called *objectives*, as are the aims of test groups.

4.2.5 Default tables

Defaults serve purposes similar to those of test steps, and, in fact, the form of a default table is quite similar to that of a test step table. Therefore, it is reasonable to present Table 4.6 here, even though we will not go into the details of default semantics until Section 4.5.

Like test steps, defaults may be parameterized. Note that defaults do not have defaults – a measure taken to contain the potential intricacies of default handling. In the 1992 version of the TTCN standard, default trees were not even allowed to

TABLE 4.5. Test step table

TEST STEP DYNAMIC BEHAVIOUR					
TEST STEP NAME : *TestStepId&ParList* – must be unique throughout the test suite GROUP : *TestStepGroupReference* – structural information OBJECTIVE : *FreeText* – an informal short description DEFAULT : [*DefaultRefList*] – a list of default references, separated by commas cf. Section 4.5 COMMENTS : [*FreeText*] – this line may perhaps also be omitted completely in the second version of the standard					
NR	LABEL	BEHAVIOUR DESCRIPTION	CONSTRAINTS REF	VERDICT	COMMENTS
:	:	:	:	:	:
m	[*Label*]	*StatementLine* (indented)	[*ConstraintReference*]	[*Verdict*]	[*FreeText*]
:	:	:	:	:	:
		TreeHeader			
n	[*Label*]	*StatementLine* (indented)	[*ConstraintReference*]	[*Verdict*]	[*FreeText*]
:	:	:	:	:	:
DETAILED COMMENTS: [*FreeText*]					

TABLE 4.6. Default table

DEFAULT DYNAMIC BEHAVIOUR					
DEFAULT NAME : *DefaultId&ParList* – must be unique throughout the test suite GROUP : *DefaultGroupReference* – structural information OBJECTIVE : *FreeText* – an informal short description COMMENTS : [*FreeText*] – this line may (perhaps) be omitted completely					
NR	LABEL	BEHAVIOUR DESCRIPTION	CONSTRAINTS REF	VERDICT	COMMENTS
:	:	:	:	:	:
m	[*Label*]	*StatementLine* (indented)	[*ConstraintReference*]	[*Verdict*]	[*FreeText*]
:	:	:	:	:	:
		TreeHeader			
n	[*Label*]	*StatementLine* (indented)	[*ConstraintReference*]	[*Verdict*]	[*FreeText*]
:	:	:	:	:	:
DETAILED COMMENTS: [*FreeText*]					

have local trees. However, TTCN user requests have led not only to the permission of default local trees, but even to new and powerful language elements to finetune default handling. This could only be achieved at the cost of some complexity in determining which defaults actually apply at any point in the expansion of the behavior tree.

As described in 4.1.5, send events must not appear in arbitrary lists of alternatives. Since defaults, on the other hand, appear potentially in expansions everywhere, it is not allowed to have a send event on a default's top list of alternatives.

Default behavior trees may attach only local trees; they must not attach test steps. The standard does not rule out create statements in defaults; however, if return statements are involved, care must be taken not to start any test component twice in the same test case.

4.2.6 Line and page continuation

In TTCN.GR tables, in particular in index and behavior tables, the problems of limited line length and limited page height arise. It may become necessary to continue a column entry on the following typographical line(s), and it may become necessary to continue a table on the following page(s). In order to avoid ambiguities, TTCN regulates line continuations in behavior table bodies and page continuation in arbitrary tables.

Column entries in behavior lines may extend over several typographical lines (let us write "t-lines", for the moment). Of course, behaviour lines do not interlace vertically; the entries of a behavior line in the various columns are written top-aligned. Expressed more formally, each entry of each new behavior line starts in the t-line below the lowest t-line used in the previous behavior line or header. TTCN offers a choice of two alternative methods of line continuation:

– Either the line number column is used to identify line continuation. This is sufficient, because each line number is written on the top t-line used for its behavior line.

– Or – no matter whether there is a line number column or not – continuation t-lines in the behavior description (and only there) are marked by an initial hash character "#".

The two methods are shown in Table 4.7. It is recommended not to use both methods in the same test suite. Line continuations in other tables are not regulated by the standard, probably because they were not expected to become necessary.

Tables extending beyond the bottom of a page are continued on the following page(s). The necessary titles – usually the column headers – are repeated on top of the continuation page. The indications of continuation shown in Figure 4.6 have to be added below the leading and above the following part of the table.

If the body of a behavior table is split between two pages, then the level of indentation of the last statement line in the previous part of the table is indicated

TABLE 4.7. Two methods of indicating line continuation

NR	L	BEHAVIOUR DESCRIPTION	CREF	VERDICT	COMMENTS
1		L! CONNECTrequest [IUT_Type = Responder]	CR1		The maximal number of t-lines used for a behavior line may appear in any column.
2		L? CONNECTconfirm	CC1		

LABEL	BEHAVIOUR DESCRIPTION	CREF	VERDICT	COMMENTS
	L! CONNECTrequest #[IUT_Type = Responder]	CR1		The maximal number of t-lines used for a behavior line may appear in any column.
	L? CONNECTconfirm	CC1		

NR	LABEL	BEHAVIOUR DESCRIPTION	CONSTRAINTS REF	VERDICT
1		!A	ConstrA	
2		?B1	ConstrB1	
3		!C	ConstrC	
4		?D1	ConstrD1	FAIL

[1] [3]

Continued on next page

Continued from previous page

[1] [3]

NR	LABEL	BEHAVIOUR DESCRIPTION	CONSTRAINTS REF	VERDICT
5		?D2 .	ConstrD2	PASS
6		?B2	ConstrB2	FAIL

FIGURE 4.6. Page continuation

below, and the level of indentation of the first statement line in the following part of the table is indicated above. If need arises, more levels will be indicated, or even a grid of all relevant levels. Since these level tags unambiguously define the indentation in the table parts, reduction of the actual typographical indentation of the statement lines by a constant space is permitted. This method of saving space by is also shown in Figure 4.6.

This technique may even solve space problems arising from excessive indentation, i.e. if a table is intentionally interrupted and the rest of the page is left blank, because indentation has moved too far. Yet, whenever a behavior table takes more than one page, it is always worth investigating whether the behavior description might not profit from the detachment of subtrees as separate test steps. Often indentation and page continuation problems will be resolved by additional modularization of the test case. Modularization will also significantly benefit readability.

4.2.7 The syntax of behavior lines

In this section, we will provide syntactical details that were hitherto omitted, and set pointers to the remaining explanations of semantics.

We recall that a "horizontal row" in the body of a behavior table may be either a local tree header or a behavior line. It is the latter case in which we are now interested. The MP equivalent for a GR behavior line is the BNF nonterminal *BehaviourLine*, whose BNF expansion was shown in 4.2.2. It contains a *Line,* which is made up of an indentation (graphical in GR, numerical in MP) and a *StatementLine*. Statement lines describe the "real actions" of a tester: sending (*Send*), receiving (*Receive, Otherwise*), checking timer expirations (*Timeout*), checking terminated PTCs (*Done*), checking values of data objects (*Qualifier*), changing values of data objects (*AssignmentList*), managing timers (*TimerOps*), expanding the current portion of the working tree (*Constructs*), or inviting the remote IUT to become active (*ImplicitSend*):

> *StatementLine* ::= (*Event* [*Qualifier*] [*AssignmentList*] [*TimerOps*]) |
> (*Qualifier* [*AssignmentList*] [*TimerOps*]) |
> (*AssignmentList* [*TimerOps*]) | *TimerOps* |
> *Construct* | *ImplicitSend*

Event lines are the most important kind of statement lines, and events are the most important elements of TTCN behavior descriptions. In terms of BNF nonterminals, we deal with:

> *Event* ::= *Send* | *Receive* | *Otherwise* | *Timeout* | *Done*

Send and receive events have quite similar forms:

> *Send** ::= ([*PCO_Identifier* | *FormalParIdentifier*] "!"
> (*ASPIdentifier* | *PDU_Identifier*)) |
> (*CP_Identifier* "!" *CM_Identifier*)

Receive* ::= ([PCO_Identifier | FormalParIdentifier] "?"
(ASPIdentifier | PDU_Identifier)) |
(CP_Identifier "?" CM_Identifier)

The FormalParIdentifier must be a test step or local tree parameter evaluating to a PCO. We are not sure whether it would have made sense to allow formal parameters for CPs, too; at any rate, it is apparently not allowed by the standard.

Otherwise events have syntactic and semantic parallels with receive events.

Otherwise ::= [PCO_Identifier | FormalParIdentifier] "?" **OTHERWISE**

Note that the otherwise event is not used for CPs and CMs – probably because its main purpose is to catch unexpected PDUs (as such or inside ASPs). The exchange of CMs, being totally in control of the test suite writer, should not have anything unexpected about it.

Timeout events are even simpler:

Timeout ::= "?" **TIMEOUT** [TimerIdentifier]

Single pseudo-events, i.e. qualifiers, assignments and timer operations, were already discussed in Subsection 4.1.6. We will fill in further semantic details in the next two sections.

AssignmentList* ::= "(" Assignment {"," Assignment} ")"
Assignment ::= DataObjectReference ":=" Expression
Qualifier ::= "[" Expression "]"
TimerOps* ::= TimerOp {"," TimerOp}
TimerOp ::= StartTimer | CancelTimer | ReadTimer

The implicit send is a language element introduced particularly for the remote test method. It is explained in 4.4.5.

ImplicitSend ::= "<" **IUT** "!" (ASPIdentifier | PDU_Identifier) ">"

Let us close this section by displaying in Figure 4.7 a typical behavior line along with its relevant BNF nonterminals.

4.3 Pseudo-events

In this section we present more details of the syntax and semantics of pseudo-events in pseudo-event lines. Their particular treatment as additions to event lines is addressed in the next section.

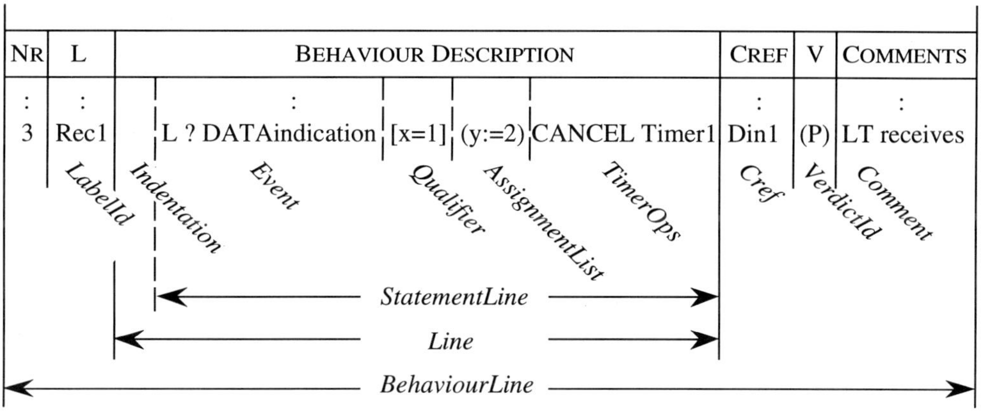

FIGURE 4.7. A behavior line and its BNF components

4.3.1 *On referencing data*

Assignments and qualifiers use expressions. Expressions may contain not only literal or ASN.1 values but also references to data objects:

- names of test suite parameters and constants, test suite and test case variables,
- formal parameters of the operation calls or trees in which they appear, the trees being either root trees of test steps or defaults, or local trees of the "current" test case, test step or default,
- names of ASPs, PDUs or CMs to be sent or received in the event line in which this reference appears, as detailed in 4.4.4.

These references to data objects may be supplemented by a list of component references, each of which starts with a dot. This list will, in the way of an access path, point to an arbitrarily deeply nested component of the referenced data object, in quite a similar way as in the reference list mechanism used in replacements, i.e. in modified ASN.1 constraints (see subsection 3.5.4, also for the non-terminals not expanded below). Unlike the case of ASN.1 replacement, it is also possible to point to components of data objects of a tabular type, i.e. to ASP or CM parameters, PDU fields and structured type elements.

*DataObjectReference** ::= *DataObjectIdentifier* { *RecordRef* | *ArrayRef* | *BitRef* }

*RecordRef** ::= "." (*ComponentIdentifier* | *PDUIdentifier* | *StructIdentifier* | *ComponentPosition*)

ComponentIdentifier ::= *ASP_ParIdentifier* | *PDU_FieldIdentifier* | *CM_ParIdentifier* | *ElemIdentifier* | *ASN1_Identifier*

Though the standard is not perfectly clear about this point, a static semantic rule for *RecordRef* seems to demand that in the access path, names of PDU fields or of structured type elements must be preceded by the appropriate PDU type name or structured type name – at least if this PDU or structured object is accessed via constraint chaining, cf. 3.5.1. Assume, for example, an ASP constraint named *ASP1* has an ASP parameter *data* that is chained to a PDU of type *PDUtype1*, and this PDU has a field *fieldA*. Then this field would be referenced somewhat redundantly as *ASP1.data.PDUtype1.fieldA,* instead of merely as *ASP1.-data.fieldA.*

One exception to the accessibility of data object components should be noted: if a PDU is declared to be of the pseudotype "PDU", single fields of this PDU must not be referenced.

When dealing with assignments or qualifiers, the standard often uses the term "(TTCN) expressions" in the natural language description. We will rather follow the nomenclature in the BNF, where *Expression* is the nonterminal for expressions in the ordinary sense, i.e. terms denoting values of a given type. Expressions are composed of operators, operations and component expressions of the appropriate types. As there are potentially many types involved, this type consistency was not reflected in the BNF, but rather prescribed by static semantics and clauses in the text. Thus, for instance, there does not exist any BNF nonterminal *BooleanExpression.*

4.3.2 Qualifiers

Qualifiers are Boolean expressions, enclosed by square brackets.

Qualifier ::= "[" *Expression* "]"

In a pseudo-event or event line, a qualifier restricts the possibility of a success (or match) of this line: independently from its other components, the behavior line does not succeed if its qualifier, evaluated before anything else, yields FALSE. Note, in particular, that a qualifier is always evaluated before any assignment on the same line is executed.

4.3.3 Assignments

Another kind of pseudo-event is the assignment list. An assignment list contains a list of one or more assignments, separated by commas, and enclosed in parentheses.

*AssignmentList** ::= "(" *Assignment* {"," *Assignment*} ")"

Assignment ::= *DataObjectReference* ":=" *Expression*

When the assignment is performed, the right-hand side must evaluate to a unique value of the same type as the data object reference on the left-hand side. As an exception to this type compatibility, different string length restrictions may be declared or evaluated for both sides. In case of inconsistency, the string to which the right-hand side evaluates is changed appropriately: if it is too long, it is truncated on the right to the maximum length declared for the left-hand side; if it is too short, it is padded on the right with fill characters, up to the maximum length declared for the left-hand side. Blanks (" ") are used to fill character strings, and zeros ("0") are used to fill BIT-, HEX- and OCTETSTRINGs.

The assignments of an assignment list are executed from left to right, and not, for example, in parallel. Thus, after (x:=y, y:=x) both x and y contain the previous value of y. The direct and indirect effects of an assignment list are, of course, affected by the order in which pseudo-events and events on the same behavior line are executed. We will deal further with this topic in the next section.

4.3.4 Timer operations

Operations for starting, reading and cancelling timers may appear on event and pseudo-event lines. In conjunction with qualifiers and or assignments, timer operations are dealt with last, i.e. they are not performed if the qualifier yields FALSE, and they respect the latest values, in particular those just changed by the assignments.

Several timer operations may be written in succession on the same line, separated by commas:

*TimerOps** ::= *TimerOp* {"," *TimerOp*}

TimerOp ::= *StartTimer* | *CancelTimer* | *ReadTimer*

A timer may be *idle* or *busy*. Moreover, there may or may not exist a timeout notification from it. Timer expirations and the execution of timeout lines and timer operations control the transitions between the following three possible timer states:

- *inactive* (idle, without timeout notification),
- *running* (busy, without timeout notification), and
- *expired* (idle, with timeout notification).

Multiple timeout notifications for the same timer will never exist at the same time, and a timeout notification will never exist for a busy timer. Therefore, the list of timeout notifications is equivalent to the current set of expired timers.

Timer operation semantics are defined in a way that avoids test case errors; even "inappropriate" applications on declared timers are interpreted in a "harmless" way, as detailed below.

STARTING TIMERS

A timer is started by a timer operation of the form

StartTimer ::= ***START*** *TimerIdentifier* ["(" *TimerValue* ")"]

with the running period defined by the predeclared time units and the positive integer duration that is defined either by the optional timer value or by the predeclared default duration. For example, the *TimerOps*

START T1, START T2(10), START T3(Var1)

start T1 with its declared default duration, T2 with a running period of 10 of the declared time units, and T3 with as many time units as the present value of Var1.

Starting a running or expired timer effects a prior cancellation: "START T1" is always equivalent to "CANCEL T1, START T1". Starting a timer takes it to the state *running*.

CANCELLING TIMERS

Cancelling a timer makes it *inactive*; busy timers return to idle, timeout notifications of expired timers are deleted, idle timers stay idle. It is possible to omit the timer identifiers as a shorthand for cancelling all timers. The syntax for a timer cancellation is

CancelTimer ::= ***CANCEL*** [*TimerIdentifier*]

READING TIMERS

The number of declared time units for which a timer has been running (since its last start) can be read into a data object by a timer operation of the form

ReadTimer ::= ***READTIMER*** *TimerIdentifier* "(" *DataObjectReference* ")"

In the final analysis, the data object must be of data type INTEGER. Idle timers (inactive or expired) yield the value 0. For example, the lines

```
?C READTIMER Timer_C (Var_C)
   + C_Tree(Var_C)
```

allow execution to branch into Tree_C upon receipt of an ASP or PDU of type C. Tree_C is then executed with Var_C set to the number of time units for which Timer_C had been running when C was received.

4.4 Event behavior lines

The done event was already treated in sufficient detail in Section 4.1.11. In this section, we consider the effect of statement lines containing a send, receive or otherwise event, but this time we include possible pseudo-events, and (except in the case of an otherwise event) constraint references in our explanation. As we are concentrating on questions

- of matching and
- of the effects on input and output queues and on timeout notifications,

labels and verdicts are still not addressed. A label does not have any impact on the success or failure, or on the effect of the single behavior line bearing it. Rather, they influence how the tester gets there and what the tester writes into the test log, as detailed later in this chapter. Verdicts affect, of course, the test result, and they can terminate the execution of a test case that would otherwise continue, but in practice this will rarely happen, as pointed out in 4.6. Due to their close connection with receive events, we will also deal with implicit sends.

4.4.1 Constraint references

The column "Constraints Ref" serves to hold references to the relevant ASP, PDU or CM constraints in Send, Receive, and Implicit Send lines. A constraint reference is written according to the following syntax:

ConsRef ::= *ConstraintIdentifier* [*ActualCrefParList*]
*ActualCrefParList** ::= "(" *ActualCrefPar* {"," *ActualCrefPar*} ")"
ActualCrefPar ::= *Value* | *DataObjectIdentifier* | *ConsRef*

As actual parameters can be parameterized constraint references, nested parentheses are possible. Thus, for instance, "DataConstr1(P1,CR1(P2))" can identify a constraint for a DATArequest-ASP, which has as its second actual parameter a CONNECTrequest-PDU to be evaluated with an actual parameter P2.

4.4.2 Abstract tester states

As pointed out before, the tester is regularly "sitting in front of a list of alternatives," from which it will determine a matching alternative. Let us call this a *major state* of the tester. These are the states mentioned in the third tree interpretation at the end of 4.1.1, where the behavior lines were identified with state transitions. Let us consider the input and output queues at PCOs, timers and the timeout list as belonging to the tester's environment; details and problems of this assumption are discussed in Section 4.9. Let us finally, leaving aside the details of constructs and of tree expansion, consider the behavior tree on which the tester is operating as completely determined, including the information given in the test suite declarations and constraints. Apart from the behavior tree, a major state of the tester is thus composed of

- the tester's position in the behavior tree (i.e. in front of which list of alternatives it is sitting), and
- the values of test suite parameters, constants, and variables, as well as test case variables. In fact we might also include the result variable here, which will be described in 4.6, but we can safely leave that aside for the moment.

The tester moves from one major state to the next, each time by executing one of the next list of alternatives. We will now fill in the last details on how the tester determines which behavior line to execute, and what happens if this line is executed. In doing so, we will consider the tester as moving through *intermediate states,* which come into play between the major states and are introduced only for the purpose of explaining operational semantics. As usual, implemented testers need not have any counterpart for these states, neither major nor intermediate.

4.4.3 Success of alternatives

In this subsection, we take a closer look at the way to determine whether or not an alternative succeeds. What finally happens during their execution, once they are successful, is the topic of the following subsection.

In a major state, if the list of alternatives is empty, a leaf of the behavior tree was just executed, and the test case terminates, as described under the topics of verdicts (4.6) and activities between test cases (4.7).

If the list of alternatives is not empty, the tester takes a snapshot of the timeout list as well as of the front positions of the input queues. We recall that the tester executes the first in the given order of alternatives to succeed, based on this snapshot. Ignoring test case errors, if none of the alternatives on the list succeeds, the tester returns the last major state; it then takes another snapshot and continues as before.

In attempting a pseudo-event, send, or timeout alternative, the tester first evaluates the qualifier, provided there is one, on the basis of the values of the parameters, constants and variables in the most recent major state. If the qualifier yields FALSE, the alternative is unsuccessful. If it yields TRUE, the remaining elements of the line are considered.

In fact if the tester has reached this point with a send line, then the alternative is successful, and the tester executes the line, as specified in the next subsection. Similarly, a pseudo-event line without a qualifier is successful. Its execution was already explained in 4.3. A timeout line without a qualifier is successful if the timeout list contains a timeout notification of the specified timer. Its execution removes either notification. Note also the "unspecific timeout" mentioned in 4.1.4. Similarly, an otherwise line without a qualifier is successful if the specified input queue contains any data object. Its execution removes the data object from its queue.

Things are slightly more complicated in the case of receive lines, the success of which is determined in the following steps:

1. Is there a data object in the specified input queue? The default PCO is specified implicitly if none is specified explicitly.

2. If the answer to question 1 is "yes", is the first element of the queue of the type specified in the receive line?

3. If the answer to question 2 is "yes", does it match the pattern defined by the constraint referenced in the receive line?

4. If the answer to question 3 is "yes", is the line without a qualifier, or, if there is one, does the qualifier evaluate to TRUE? This evaluation uses the values of the data object to be received, provided any of its components are referenced!

As soon as any one of these questions is answered negatively, the receive line is unsuccessful. If all answers are "yes", the line succeeds.

The *anticipatory evaluation* of a receive qualifier permits basing the tester's decisions on details of the received data object. It is unnecessary to store it first and to decide later on the basis of the stored value. Of course, appropriately defined constraints, for instance modified or parameterized, could often be used instead. However, the anticipatory evaluation mechanism may save some constraint declarations, and, what is more important, permits the formulation of complex conditions that cannot be expressed with only the language elements provided for constraints.

4.4.4 The execution of send or receive lines

The execution of a successful send or receive line generally affects simultaneously:

- the major state of the tester:
  - By the execution of an event or pseudo-event line, the position in the behavior tree changes; the next list of alternatives, written in TTCN trees below the executed line at the next higher level of indentation, is addressed.
  - The performance of assignments and timer readings may change values of variables, as described before. Likewise, a preliminary verdict may change the value of the result variable, cf. 4.6.
- its environment, consisting of timers, the timeout list, input and output queues:
  - An executed receive line removes the matched data object from the front of the specified receive queue. The parts of the data object needed later in the test case have to be saved by explicit assignments.
  - An executed send line appends the data object defined by the constraint to the end of the specified output queue.
  - Timer operations on the executed line alter timer states, as previously described.
- the test log:
  - As a minimum requirement, the labels, events and verdicts of all executed behavior lines are recorded.

While qualifiers play a special role in the matching attempts of receive lines, assignments play a special role in the execution of send lines. In receive events, supplementary assignments are performed only *after* the respective event was matched and performed, i.e. taken from the queue. This makes a difference if, for example, the receive constraint has a variable as an actual parameter and this variable is assigned a new value in the assignment. In the context of timeout and otherwise lines, it does not matter whether the assignments are performed before or after the event. In the execution of a send line, the assignments on the line are performed *before* the data object defined by the constraint is put into the specified output queue. In particular, it is possible to make "last minute" assignments to particular ASP or CM parameters or PDU fields, which override the values found in the send constraint.

The two special cases can be summarized by the following rule of thumb:

- Anticipatory qualifiers in receive lines
- Last minute assignments in send lines

Let us illustrate anticipatory qualifiers, last minute assignments, and the storing of received values by means of a very simple example. Assume *A* is an ASP type with two parameters, *Par1* and *Par2*. Par1 is of type INTEGER, and Par2 is of type BOOLEAN. Then the receive statement

? A [A.Par2]

with constraint *A_Constr* is only successful if there is an A at the front of the input queue at the PCO, if this A matches the constraint A_Constr, and if the second parameter of A is TRUE. The last condition is an example of the anticipatory qualifier effect. If the receive statement is not successful, A is still in the queue, and, in fact, neither the tester's variables nor the environment have really changed.

If the tester were to send an A by the statement line

! A (A.Par1=0)

with constraint A_Constr, then the A would be sent with a first parameter value of 0, irrespective of the value given in the first parameter of A_Constr. This is the effect of the last minute assignment.

If the tester were supposed to store the values of both parameters of a received A, this might, for instance, be accomplished by means of two variables, *V_Int* and *V_Bool,* and the following receive statement:

? A (V_Int := A.Par1, V_Bool := A.Par2)

The relationship between qualifiers, assignments, sends and receives can be illustrated by the following symbolic and very informal "expansions," that can only serve as mnemonic aids, but not as actual expansions in the working tree. With these reservations, a complex send line with a next alternative and a sequentially following line,

```
send event   qualifier   assignment list   timer operations
    etc. (subsequent behavior)
etc. (alternative behavior)
```

could be viewed as

```
qualifier
    copy the data object defined by the constraint
        assignment list (where applicable, applied to the copy)
            send event (using the, possibly modified, copy)
                discard copy
                    timer operations
                        etc. (subsequent behavior)
etc. (alternative behavior)
```

With the same reservations, a complex receive line with an alternative and a sequentially following line,

```
receive event/otherwise  qualifier  assignment list  timer operations
    etc. (subsequent behavior)
etc. (alternative behavior)
```

could be viewed as

```
copy the front event
    qualifier (evaluated on the base of this copy)
        receive event/otherwise
            assignment list
                timer operations
                    etc. (subsequent behavior)
    discard the copy
        etc. (alternative behavior)
etc. (alternative behavior)
```

4.4.5 Calling for something to be received: the implicit send

In the remote test method it is often assumed that a local SUT operator can exert the same influence on the IUT as an upper tester at a PCO on the IUT's upper boundary. This refers chiefly to

- the choice between alternative reactions, such as accepting or rejecting a request of the lower tester to set up a connection, and
- the initiative for active behavior, such as the IUT's producing a request to set up a connection,

both of which are usually effected by the user entity above the IUT. The implicit send causes this behavior to be initiated when no upper tester exists. By a TTCN statement of the form

ImplicitSend ::= "<" ***IUT*** "!" (*ASP_Identifier* | *PDU_Identifier*) ">"

the test case tells the SUT operator: "Please make the IUT send the specified ASP or PDU towards the lower tester." It is only allowed to call for IUT send actions that could be enforced by a user of the IUT.

An implicit send is never accompanied by pseudo-events or a verdict, but always accompanied by a constraint reference, telling what is to be sent.

For each implicit send, the PIXIT form contains the question whether the client is able to control the IUT in the required way. A test campaign using the remote method would, of course, be of limited value if all these questions were answered negatively.

Typically, the implicit send is written just ahead of the corresponding receive line, which contains the same constraint reference, e.g. as in

BEHAVIOUR DESCRIPTION	CREF
<IUT ! CR>	CR1
LT ? CR	CR1

which looks, in a way, quite redundant.

4.5 Constructs and defaults

As previously noted, a *StatementLine* may be a *Construct*. Therefore, construct lines never contain pseudo-events. A construct is either a jump, an attachment, a loop, a default control command, or a create:

Construct ::= *GoTo* | *Attach* | *Repeat* | *Return* | *Activate* | *No_Defaults*
| *Create*

Constructs in TTCN trees are used to build up the relevant parts of the behavior tree, i.e. the working tree, during the execution of a test case. They can help

- to save space in the representation of large behavior trees,
- to represent infinite behavior trees by a finite description, and
- to structure test cases clearly.

The most general construct in non-concurrent TTCN is tree attachment. Jumps and loops can be regarded as variants of tree attachment. If tree parameterization is used in addition to the modularization rendered possible by the use of constructs, space savings become even more significant.

The default mechanism is a short way to represent multiple tree attachments. By the recent addition of activate, return and no_defaults constructs [ISO 9646-3-1], it has become more flexible, but also more complicated.

The create construct is used to start parallel test components. Unlike the other constructs, it does not contribute to the growth of the single working tree used in non-concurrent TTCN. A create generates a new working tree, which is executed concurrently with the previously active working trees.

As noted before, for each test case we deal with three kinds of trees:

- TTCN trees are the trees in the behavior description: root trees or local trees of the test case, of test steps, or of defaults. They are all finite and usually contain constructs.
- The behavior tree is the usually infinite tree produced conceptually by repeatedly expanding all constructs in the test case root tree.
- The working trees are built up dynamically, as described above. They are all finite and may still contain constructs after the present position of test execution.

As long as the next list of alternatives still contains one or more constructs, these constructs have to be expanded. The expansion of a construct will frequently introduce other constructs into the list, for example if a tree is attached which contains attach lines among the first level of alternatives. In order to guarantee that a finite number of expansion steps suffices for any list of alternatives with constructs, some suitable restrictions were introduced, which sometimes slightly complicate the otherwise relatively simple rules for constructs. Other rules were introduced to prevent the introduction of unreachable lines in the working tree, though there is not always a compelling reason to do so. We will present these rules in our treatment of the constructs.

Numerous examples will be used in this section to illustrate the effects of constructs. For the sake of simplicity, they will be written in a kind of pseudo-TTCN, mostly without table or column headers or boundary lines, omitting irrelevant columns, and without distinguishing between root trees and local trees.

4.5.1 Attach statements

The attach statement points either to a test step in the test step library or to a local tree in the same test case, test step or default table. Here, "the same" denotes the table in which the TTCN tree containing the attach line is specified. Attached trees may be parameterized.

Attach ::= "+" *TreeReference*[*ActualParList*]

TreeReference ::= *TestStepIdentifier* | *TreeIdentifier*

The various attachment possibilities were presented in 4.1.9.

SUBSEQUENT BEHAVIOR

One technical aspect that was left out of consideration in the preliminary treatment in 4.1.9, was the question of subsequent behavior after an attach line. In fact, an attach line can be followed by additional behavior lines. During working tree expansion, these lines are attached to all leaves of the attached tree, when this tree replaces the attach line. Thus, the tree …

```
Tree1                 Tree2                          Tree1
A          with …     C             amounts          A
  +Tree2                D           to the tree …      C
    B                 E                                  D
                                                           B
                                                       E
                                                         B
```

Still, Tree2 is called a subtree of Tree1, even though it does not extend down to the leaves of Tree1.

Other important points of tree attachment semantics concern parameter replacement and recursion.

PARAMETER REPLACEMENT

The replacement of the formal parameters of an attached tree by the actual parameters of the attachment "call" is defined as textual substitution, with two different scopes, depending on the kind of tree to be attached:

- In the attachment of a local tree, the formal parameter occurrences in this single TTCN tree are replaced by actual parameters.
- In the attachment of a test step, the formal parameter occurrences are replaced by actual parameters not only in the test step's root tree, but also in the local trees of this test step.

  Though it is not explicitly stated in the text of the standard, both an example in the standard and the above rule suggest that the formal parameters of a test step are not only formal parameters of its root tree, but also of its local trees. There, they are not declared in the header, but may be used in the body. Formal parameters declared in the headers of local trees of test steps are additional parameters, joining those inherited from the test step table.

  Without such rules, the textual replacement called for would in some cases produce nonsensical results.

Even the simpler of the two cases, the first one, gives rise to problematic examples. If the local tree Tree2 is specified as

```
Tree2(PCO:TSAP; X,Y:INTEGER)
PCO?CONind (X:=Y)
```

then the tree

```
Tree1(L:NSAP; U:TSAP)
L!CONreq (M:=1)
   +Tree2(U,M,2)
```

is equivalent to

```
Tree1(L:NSAP; U:TSAP)
L!CONreq (M:=1)
   U?CONind(M:=2)
```

Obviously, though not explicitly demanded by the standard, formal and actual parameters of test steps should be used in such a way that only valid TTCN is created by textual substitution. With some actual parameters, parameter substitution offers unpleasant surprises. The above example is taken almost verbatim from the standard. With minute changes, it turns into a counter-example. For instance, one should not write "+Tree2(U, M+1, 2)," because then textual replacement would create the invalid assignment "M+1:=2." We recommend using formal parameters in positions where only their value is "read."

RECURSION

Tree attachment may be specified iteratively, in a nested way, such that attached trees themselves attach other trees; it is perfectly valid to come back to the "outermost" tree via a recursive cycle of several attachments. One TTCN tree may even attach itself directly.

Thus, the tree ...

```
Tree1                    Tree2             amounts          Tree1
A            with ...    C                 to the           A
   +Tree2                   +Tree1         equally valid       C
                         D                 tree ...               +Tree1
                                                               D
```

Both versions of Tree1, if fully expanded, lead to the same infinite behavior tree, whose first 4 levels of alternatives look like the schema shown in Table 4.8.

Infinite behavior trees are not necessarily undesirable. If they include timeouts or counter mechanisms in the right places, the test case will terminate properly. In the preceding example, the test case will terminate if care is taken to ensure that sooner or later C is unsuccessful, but D succeeds. As shown in the following subsections, other constructs, expressed in terms of attachment, lead quite naturally to recursion.

TABLE 4.8. An infinite behavior tree

```
Tree1
A
    C
        A
            C
                ...
            D
    D
```

In one respect, however, care is taken to avoid infinite regress: it is conceived that the tester will want to expand in a simple manner all constructs in the next list of alternatives. This would, at least in the simple operational semantics used, be prevented by a tree's recursively attaching itself on the top level of alternatives, be it directly or indirectly. An expansion step of the tree …

```
Tree1                 Tree2                       Tree1
A          with ...   C          would yield ...  A
+Tree2                    D                       C
                      +Tree1                          D
                                                  +Tree1
```

and would still leave an attachment to resolve on the top level. Any number of further attachment expansions could not change this. Therefore, a tree cannot attach itself in the top list of alternatives, even indirectly. Of course, by reachability arguments (no line below the first three lines in the expanded Tree1 could ever be reached) it is at least occasionally possible to identify situations where expansion might easily be stopped. Perhaps the committees defining TTCN did not want to impose subtle considerations of that kind on implementors.

4.5.2 Goto statements

Conceptually, the *goto* statement redirects the position of the tester in the behavior tree back to an ancestor list of alternatives, one or more levels higher up than the goto. As with any other construct, there is no notion of an unsuccessful goto. If anything prevents the execution of the goto, such as the absence of a label of the required name, it constitutes a test case error. A goto has the form

GoTo ::= ("->" | ***GOTO***) *Label*

In the tree in Table 4.9, the goto statement points back to the list of alternatives consisting of B and D.

TABLE 4.9. Valid and invalid target labels of gotos

LABEL	BEHAVIOUR DESCRIPTION
LA	A
LB	B
LC	C
LD	D
LE	E
LF	GOTO LB

The target label in the goto statement must be the label of the first alternative in the target list. In the last line of Table 4.9, GOTO LD (instead of LB) would be invalid, because it would point to the second alternative of the list (B,D). To state this in a different way: a goto does not necessarily lead back to the execution of an ancestor line in the tree; rather, it effects a jump into a position, i.e. a major state, *after* some ancestor of the goto line. The valid target labels from the goto line are LA, LB and LE.

In fact, LA is only valid if the TTCN tree shown is the root tree of a test case, because goto jumps to the top list of alternatives are only allowed in the test case root tree. The latter restriction has the following rationale: test steps, defaults, and local trees may be attached in positions other than at the beginning of a list of alternatives. In this case, upon expansion, a goto target label associated with the first of the attached tree's top level of alternatives would end up pointing at an alternative that is not the first in its list.

Note that goto jumps and the unique meaning of labels are limited to one and the same TTCN tree. There are no goto jumps from one TTCN tree to another. Conditional jumps are achieved by letting goto lines follow sequentially the appropriate qualifier lines.

In a way, any goto could be dispensed with, because the same effect can be achieved by using tree attachment. Table 4.10 shows an equivalent way of writing the previous example. Labels were omitted for the sake of simplicity, but could be included for logging purposes.

The TRUE is not necessary in this particular example, but in general, if the goto has other alternatives following it in the list, the TRUE is necessary to make sure that the top level of the newly created subtree is the only way to continue after the jump. On the other hand, due to the certain success of the goto, it is unwise to have it anywhere in a TTCN tree except in a leaf position and as a last alternative. Later alternatives and subsequent behaviour would be unreachable lines.

TABLE 4.10. A goto replaced by tree attachment

```
BEHAVIOUR DESCRIPTION
A
    +BD_Tree

BD_Tree
B
    C
D
    E
        [TRUE]
            +BD_Tree
```

Another advantage of replacing gotos by tree attachments, at least as far as operational semantics is concerned, is that it avoids building a behavior tree or working tree with ambiguous goto targets, which happens if different TTCN trees with identical labels are glued together. Note, finally, that replacing gotos by attachments always produces attachment recursion.

4.5.3 Repeat statements

By using a loop in TTCN, specified by a *repeat* statement, test suite writers can cause a test step or a local tree to be executed at least once, and then as often as necessary to satisfy a particular exit condition. The general form of repeat statements is described by the following BNF rule:

Repeat ::= ***REPEAT*** *TreeReference* [*ActualParList*] ***UNTIL*** *Qualifier*

TreeReference identifies the tree to be repeated, if applicable with the desired actual parameters in *ActualParList*. *Qualifier* denotes the exit condition.

Assume, any number and kinds of ASPs, for example, are to be received at a PCO called L, until an ASP of type B has been received. This is achieved by the repeat statement in

```
(it_was_a_B:=FALSE)
    REPEAT ReceivingSomething UNTIL [it_was_a_B]
```

referring to the test step or local tree

```
ReceivingSomething
L?B (it_was_a_B:= TRUE)
L?OTHERWISE
```

The general effect of the repeat is easily explained by transforming it into an equivalent piece of TTCN tree with a tree attachment and a goto. Assuming the label LAB is not used in the TTCN tree containing the repeat, the lines

```
REPEAT SomeTree(Parm1,...,ParmN) UNTIL [Condition]
    etc. (subsequent behavior)
```

can be replaced by

```
            [TRUE]
LAB             +SomeTree(Parm1,...,ParmN)
                    [NOT Condition]
                        -> LAB
                    [Condition]
                        etc. (subsequent behavior)
```

As shown above, subsequent behavior would have to be indented two additional units, so as to make it subsequent to the final condition. In fact this final condition could be left out, but it makes the expansion easier to understand. The [TRUE] on top is used to avoid both invalid TTCN and undesired semantics: without this dummy line, the expansion of a repeat that is not a first alternative would yield a goto target label that is not associated with a first alternative. Moreover, without this line, if the tree attachment were unsuccessful, the alternatives following the repeat would be tried, and one of these might succeed, in contradiction to the guaranteed success of any repeat.
The expansion of labeled constructs is not treated specifically in 9646, but it is not ruled out, either. We interpret the label to go with the first alternative of the top list of alternatives in the expansion of any labeled construct line.

As shown before, the goto in the expansion of the repeat could also be replaced by an attachment construction, such that not only the goto, but also the repeat may be considered as a particular form of tree attachment.

Test suite writers who prefer while type loops, in which the exit condition is tested before the action, may use qualifiers and indentation to “program” these loops. Do type loops, in which the exit is controlled via the number of repetitions, can be realized as special cases of while loops. The rules for the TTCN constructs prevent the writing of loops with jumps right into or out of the middle of the loop body.

4.5.4 Table-wide defaults

In this subsection we deal with the defaults as they were described in the 1992 first version of TTCN. A subsequent version of TTCN will introduce additional mechanisms, which were included at the explicit request of test suite writers and which we will present in the next subsection. It will also dispense with the original restriction that defaults must not have local trees.

We have already shown that we can reduce the size of TTCN trees by replacing identical sublists of alternatives and their identical subsequent behavior by attach statements and a definition of the subtree to be attached. This can happen at both the level of a behavior table, by introducing a local tree, and on a test suite global level, by introducing a test step. It amounts to single explicit tree attachments. A default is a shorthand for many attachments of the same TTCN tree. It amounts to multiple implicit tree attachments.

Assume a subtree appears not only in a few scattered places, but regularly at the end of each list of alternatives in a test case or test step behavior table. TTCN permits omitting even the attach statements replacing this subtree. For this purpose, the subtree definition is not given as a local tree or test step, but as a *default table*, and a reference to this default table is given in the header of the test case or test step table. The form of the default tables, similar to that of test steps, was already described in Subsection 4.2.5. Figure 4.8 shows a schematic example of the use of defaults and the additional savings it permits, as compared with attachment.

The syntax of default references is a list of "tree calls;" the identifiers of the trees must be the names of default tables:

DefaultsRef ::= ***$DefaultsRef*** [*DefaultRefList*]
*DefaultRefList** ::= *DefaultReference* {"," *DefaultReference*}
DefaultReference ::= *DefaultIdentifier*[*ActualParList*]

1992 TTCN permitted only a single default reference, instead of a list. In fact, as long as we stick to table-wide defaults, this does not make much difference, because the trees in the list are to be tried in the order shown. Naming, for example, Def1, Def2 as defaults amounts to using a default

Def_1_and_2
+Def1
+Def2

Therefore, within this subsection we only speak of single defaults.

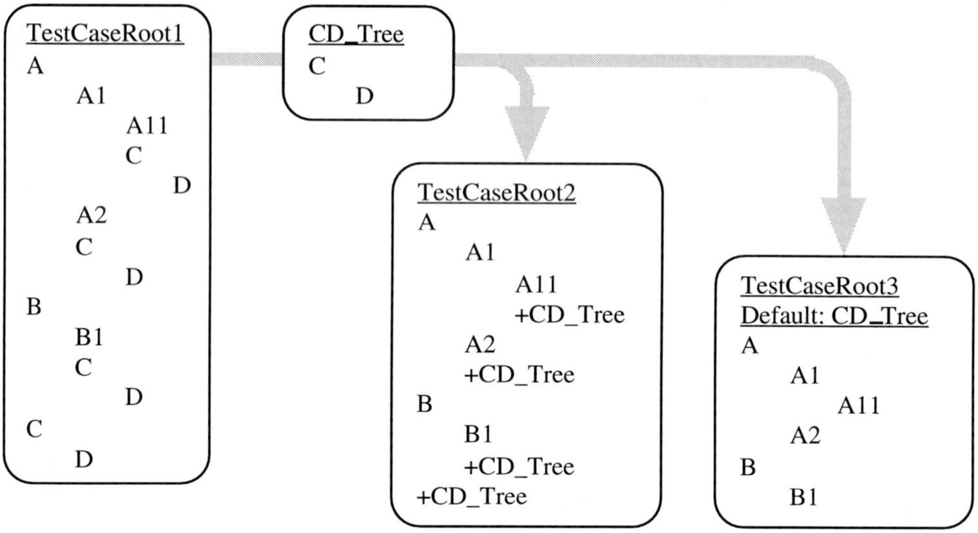

FIGURE 4.8. The default mechanism, compared to explicit attachment

Even if a subtree is not exactly identical to the last part of all alternative lists in the behavior table, but only to a considerable number of them, the default mechanism may still be used. One possibility is to restructure the test case, i.e. to divide the behavior table under consideration into pieces. Thus, one can take apart TTCN trees by introducing local trees and attachment lines, and one can turn some of the local trees into test steps, each with a different default. Another possibility is to use the dynamic default control statements presented in the next subsection.

Defaults are only possible for test case and test step tables, but not for default tables. Defaults identified in the headers of the former tables must be declared in default tables; it is, for instance, not permitted to write a test step name into the default entry of a behavior table.

TTCN rules that, in the construction of the behavior or working tree, defaults are not added to the top list of alternatives, except in the case of the test case root tree. A useful rule of thumb is "The first level of alternatives gets its defaults from the calling tree." Hence, if the rightmost tree in Figure 4.8 were not a test case root tree, its expansion would only yield the leftmost tree minus the last two lines. The standard states as the reason for this restriction that otherwise the same defaults might be expanded into several positions of the top list of alternatives. This reason is not very convincing, because it will happen often enough that defaults expand unnecessarily, for instance into alternatives to constructs or send lines. At any rate, if the default is intended to apply to the top list of alternatives for some tree, other than a test case root, it has to be appended explicitly, either written out as the full subtree or attached under a new name, under which it is defined as a test step or a local tree.

If a list of alternatives is to be expanded to replace constructs and to append defaults, ambiguities would arise from different orders of expansion. In order to eliminate these, TTCN operational semantics calls for the expansion of defaults before the expansion of constructs. In particular, if one tree attaches another tree, then this attachment is performed only after the default tree has been added in all the necessary places in both trees.

Two examples will show the effect of this rule. In Table 4.11, the left-hand side shows three TTCN trees from three different behavior tables: TopTree must be in a test case or test step table, because it attaches a test step; StepTree must be from a different test step table, because it is attached and has defaults; Step-Default must be defined in a default table. The TopTree on the left expands into the tree on the right. The middle tree shows the test case after the expansion of the default. This example shows the effect of the subtree's defaults being expanded before the subtree is attached. Note first that the E does not go with the B, because SubTree is not a test case root, and second that the default E does not carry over to the level of the D's.

TABLE 4.11. Default expansion in a tree to be attached

```
TTCN trees:                 partly expanded:          fully expanded:
----------------------------------------------------------------------
TopTree                     TopTree                   TopTree
A                           A                         A
    +StepTree                   +SubTree                  B
        D                           D                         C
                                                                  D
StepTree                    SubTree                           E
Default: StepDefault        B                                     D
B                               C
    C                           E

StepDefault
E
```

Table 4.12 has a similar structure. This example shows the effect of the top tree's defaults being expanded before the subtree is attached. TC_Root is assumed to be a test case root tree. For similar reasons as before, the three TTCN trees are taken from three different behavior tables. Note that the default E does not carry over to the level of the C.

TABLE 4.12. Default expansion in a tree attaching another one

TTCN trees:	partly expanded:	fully expanded:
TC_Root	TC_Root	TC_Root
Default: RtDefault	A	A
A	+StepTree	B
+StepTree	E	C
	E	E
StepTree		E
B	StepTree	
C	B	
	C	
RtDefault		
E		

An important application of defaults consists in the factoring out of uninteresting common alternatives, like routine otherwise and timeout statements with the subsequent actions necessary to terminate the test case cleanly, serving as backup alternatives for unsuccessful receive lines.

The standard prescribes terminating the test case forcefully in the leaves of all default trees by assigning final verdicts to them. This refers to default trees expanded by local trees of the default table, wherever applicable. Thus, in Table 4.11, the E in the fully expanded TopTree would terminate the test case, and the D after the E is automatically an unreachable line.

In principle, both defaults and constructs are very useful in helping to structure and to decrease the size of test cases. However, due to their sometimes slightly tricky technical implications, care must be taken in their application. We strongly advise against the arbitrary mixing of defaults and constructs, because it may lead to tangled test cases and unintended test behaviour. As a rule of thumb, defaults should be applied like spice; they develop their best effect if used sparingly.

4.5.5 Fine-grained defaults

We recall from 4.1.9 and 4.4.2 that the tester can be thought of as

- either moving down the fully expanded conceptual behavior tree from major state to major state
- or moving down a continually changing hybrid *working tree* that is virtually a slender subtree of the behavior tree with some unexpanded constructs still in it. However, on the path the tester has already traversed, all constructs have been expanded, because the tester will only take the next action after the list of alternatives for this action has been fully expanded.

At any rate, in the first IS version of TTCN, operational semantics were not much concerned with defaults, because the latter were assumed to be previously expanded into explicit attachments in every TTCN tree attached to the working tree. In the second version of TTCN, this will change. Default behavior is now completely kept out of the working tree, but the information about the currently applicable defaults is retained as a new component of the tester's state: the *default context*. The default context consists of the currently relevant list of default references, pointing at the default behaviors that apply to the next list of alternatives. By an appropriate handling of the default context, the effect of the pre-attached tablewide defaults described in the previous subsection is easily replicated.

TTCN Extensions [ISO 9646-3-1] will introduce two new constructs for dynamically controlling the defaults, such that the default context can be changed even inside behavior tables: these are activate and return statements. Their syntax is simple:

Activate ::= ***ACTIVATE*** "(" [*DefaultRefList*] ")"

Return ::= ***RETURN***

An *activate* replaces the present default context by the specified list of default tree references. Thus, the default context is no longer necessarily constant throughout an entire behavior table; activates within the table may override it. As the default list may be empty, it is possible to switch off all presently applicable defaults by saying ACTIVATE( ).

The conventional specification of a tablewide default context by a default entry in the behavior table header, which is still valid without reservations, could now be replaced by specifying an empty default list in the header and adding activate statements referencing the desired default list in the root tree of the table:

- for test cases, it is equivalent to a single activate, prefixed to the root tree as a top single alternative;
- for test steps, it is equivalent to activate statements inserted before all behaviors following the top list of alternatives of the root tree. This complication stems from the non-applicability of table header defaults to the top list of alternatives, outside of test case roots. Additionally, behavior subsequent to attachments of test steps (cf. 4.5.1) must be prefixed by activate statements re-introducing the default context of the attaching tree.

The resulting default expansions are the same, regardless of whether the desired defaults are specified in the header or in activate lines.

Though it is not explicitly mentioned in 9646, we interpret the default mechanics to imply that the active default context carries over dynamically to attached local trees and is not overridden by the default list given statically in the table header.

Let us illustrate the handling of the default context by an example that merges features of the examples in Tables 4.11 and 4.12. Table 4.13 shows the possibility of replacing tablewide defaults by activate statements.

TABLE 4.13. Tablewide defaults and their simulation by activate lines

Defaults via headers:	Defaults via constructs:
TC_Root	TC_Root
Default: RtDf	ACTIVATE(RtDf)
A	A
+T_Step	+T_Step
D	ACTIVATE(RtDf)
	D
T_Step	T_Step
Default: StDf1,StDf2	B
B	ACTIVATE(StDf1,StDf2)
C	C

Defaults in both cases:

StDf1	StDf2	RtDf
E1	E2	E

Table 4.14 shows the behavior tree resulting for both alternatives. The root tree's default E does not carry over to the attached step tree, nor does the step

tree's defaults E1 and E2 carry over to the subsequent behavior D. The 1994 version of the TTCN standard will contain several more elaborate examples demonstrating the handling of the default context.

TABLE 4.14. Full expansion of Table 4.13

```
TC_Root
A
   B
      C
         D
         E
      E1
         D
         E
      E2
         D
         E
   E
E
```

A *return* effects a jump out of default behavior, back to the top of the list of alternatives in a test case or test step to which the currently active default behavior was appended. As a jump command, it is similar to the goto, except for the following differences:

- a goto jump does not leave the original TTCN tree and has a statically predetermined target, while a return jump always leaves the original (default) TTCN tree towards a target that is determined by the particular test case execution history;
- a return may appear on the top list of alternatives of a default root tree, even though that would usually not make much sense; a goto must not appear there.

We illustrate the effect of the return by way of the example in Table 4.15. In fact, to take care even of the case of an (improbable) top level return, a [TRUE] should always precede the goto in the expansion of the return. We showed only the simpler replacement possible in all other cases.

Another new default control feature that will probably be introduced in [ISO 96463-1] is the *no-defaults* construct. At the time of writing this text, it consists of a line "NO_DEFAULTS," followed sequentially by a list of alternatives. Its effect is to deactivate the present default context exactly for this list of alternatives. The default context will apply again for subsequent behavior. For example, if DEF1, DEF2 is the present default context, then

```
NO_DEFAULTS
    A
    B
        C
```

is equivalent to

```
ACTIVATE()
    A
    B
        ACTIVATE(DEF1,DEF2)
            C
```

TABLE 4.15. The return construct

Application of return			Expansion:		
TC_Root			TC_Root		
Default: RootDf					
A		LA	A		
	B	LB		B	
				C	
RootDf					->LB
C				D	
	RETURN		C		
D				->LA	
			D		

4.6 Verdicts

4.6.1 Verdicts in non-concurrent TTCN

In the verdict column,

- preliminary verdicts (in parentheses),
- final verdicts (without parentheses), and
- the result command R

can be written. In 9646, the result command, as it is called here, is classed with the final verdicts and is called the result variable. Various notations are allowed for verdicts:

Verdict ::= *Pass* | *Fail* | *Inconclusive* | *Result*
Pass ::= ***PASS*** | ***P*** | "(" ***PASS*** ")" | "(" ***P*** ")"
Fail ::= ***FAIL*** | ***F*** | "(" ***FAIL*** ")" | "(" ***F*** ")"
Inconclusive ::= ***INCONC*** | ***I*** | "(" ***INCONC*** ")" | "(" ***I*** ")"
Result ::= ***R***

In every test suite, a reserved test case variable R of the predefined type R_Type is declared implicitly, having the possible values *pass*, *inconc*, *fail* and an initial value of *none*. R may be "read," for example in the pseudo-events [R=fail] and (X:=R), but not explicitly "written to," as in the invalid pseudo-event (R:=pass). The value of R may only be changed by *preliminary verdicts* as follows: if the possible values of R are considered as ordered by their "degree of disapproval," i.e. none < pass < inconc < fail, then the execution of a behavior line with a preliminary verdict assigns to R a new value that is the maximum of the preliminary verdict in the behavior line and the old value of R. In other words, a preliminary verdict can only change for the worse.

If a line with the *result command* "R" is executed, the current value of R becomes the final verdict of this test case, unless it is "none", which would constitute a test case error. After this line, the test case terminates; subsequent behavior is never reached.

An explicit *final verdict* "PASS" or "P", etc., overrides R and also terminates the test case, unless the explicit final verdict is less negative than R. The latter case constitutes a test case error.

Table 4.16 summarizes in the left half how R is computed in the case of a preliminary verdict and in the right half how the final verdict is computed in the case of a result command or an explicit final verdict. As indicated in the BNF, PASS, INCONC, and FAIL could be abbreviated as P, I, and F, respectively. For example, if R=inconc, and a line with "(FAIL)" is executed, then R changes to fail. If R=inconc and a line with "R" is executed, then the test case terminates with the verdict "inconc."

Verdicts are most in order when the tester acquires new information about the IUT's conformance, i.e. in receive, otherwise or timeout lines. Verdicts are explicitly disallowed in construct lines and implicit sends. In otherwise lines, a PASS is not allowed and a FAIL is recommended. All leaves of the behavior tree without any verdict entry are implicitly treated as if they contained a result command "R". Leaves of default behavior, even if subsequent behavior is specified to come after them, are required to have a final verdict or result command

TABLE 4.16. The computation of verdicts

Old value of R	New value of R resulting from the old R and a verdict column entry of			Test case verdict resulting from the old R and a verdict column entry of			
	(PASS)	(INCONC)	(FAIL)	PASS	INCONC	FAIL	R
none	pass	inconc	fail	pass	inconc	fail	(error)
pass	pass	inconc	fail	pass	inconc	fail	pass
inconc	inconc	inconc	fail	(error)	inconc	fail	inconc
fail	fail	fail	fail	(error)	(error)	fail	fail

entry. In fact, the description of default semantics might have been simpler had writing behavior subsequent to attach lines in tables with defaults (cf. Table 4.11) been prohibited.

4.6.2 Verdicts in concurrent TTCN

In concurrent TTCN, each test component has its own private result variable. Its name is R in the PTCs and MTC_R in the MTC. A *private result variable* can be read in the test component and is affected by preliminary verdicts in the test component in exactly the same way as the single result variable can be read and is affected by preliminary verdicts in non-concurrent TTCN. Each activated PTC must contribute at least a preliminary or final verdict, or a final result. The effect of these on the termination of the PTC is the same as in non-concurrent TTCN.

Besides the private result variables, there is also a *global result variable R* that can only be read in the MTC. All final results, final verdicts and preliminary verdicts in PTCs, as well as all preliminary verdicts in the MTC, are treated like preliminary verdicts contributing to the global R, in the way described in the previous subsection, in Table 4.16. Thus, as things "can only get worse," the global R value can always only be the same as, or more negative than, any local R or the MTC_R value. Note that the transfer of "local" verdicts to the global R need not be specified explicitly by means of CMs; it happens automatically. Of course, the MOT will have to do the work, but at least the ATS writers need not do it.

The MTC is in charge of assigning the final verdict, either by assigning an explicit final verdict, or by an explicit or implicit result command. The MTC's result command will use the value of the global R as the test case verdict.

The terminating effect of verdict entries is, on the level of a test component, the same as in non-concurrent TTCN on the tester level. If any test component is to be informed of the value of any result variable other than its own, the value must be transmitted to it explicitly via CMs by the unique component that can "read" this result variable.

4.6.3 Verdict problems

At this point we want to take a closer look at the exact wording of the CTMF definition of the test case verdicts, which we therefore reproduce here:

"*pass (verdict)*: A test verdict given when the observed test outcome gives evidence of conformance to the conformance requirement(s) on which the test purpose(s) of the test case is (are) focused, and when all test events are valid with respect to the relevant specification(s).

fail (verdict): A test verdict given when the observed test outcome either demonstrates nonconformance with respect to (at least one of) the conformance requirement(s) on which the test purpose(s) of the test case is (are) focused, or contains at least one invalid test event, with respect to the relevant specifications.

inconclusive (verdict): A test verdict given when the observed test outcome is such that neither a pass nor a fail verdict can be given."

This sounds very carefully and precisely worded, almost in the style of legal documents. Nevertheless, the bearing of these definitions on test purposes and test cases depends crucially on the interpretation of some central terms and phrases:

1. When is a test event valid?
2. What is a conformance requirement?
3. What is a test purpose?
4. Which are the conformance requirements a test purpose focuses on?
5. How is evidence of conformance to a conformance requirement given?
6. How is nonconformance to a conformance requirement demonstrated?

The standard gives only informal circumscriptions of some of these notions. Let us, therefore, attempt a more elaborate, but still relatively informal, analysis of the above definitions, in order to derive an interpretation of the test case verdicts. We are quite certain that this interpretation will not meet the unanimous approval of all experts – nor will, to our knowledge, any other existing interpretation. Of course, the consequence should be to invest further standardization efforts into an unambiguous determination of test case verdicts. Without indisputable test verdicts, conformance testing would be quite meaningless. So, here is our attempt. It is confined to non-concurrent observations, i.e. it serves only for non-concurrent TTCN. However, a comparable analysis for concurrent observations requires nothing in addition but the relatively straightforward concept of a set of behavior trees with one root tree among them that starts first and triggers the others, reflecting the properties of the MTC and the PTCs.

An OSI specification defines a set of valid behaviours for its implementations. Let us call the system consisting of an IUT and the underlying provider an *IUT+*. Due to the standardized test methods and configurations, it is the *IUT+* which is being observed and controlled in a test case run. Hence the specification *Spec* of the IUT and that of the underlying provider have to be considered together as one single specification *Spec+* of the *IUT+*. We consider any parameters of *Spec+* as already replaced by the values given in the ICS and IXIT, such that *Spec+* is not parameterized.

The behavior of the *IUT+* is observed by behavior trees, which usually not only observe, but also influence this behavior. We assume everywhere that they are compatible with the *IUT+*, in terms of working with the proper PCOs. An *observation* of an *IUT+* and a behavior tree *BT* is a sequence of observation events, i.e. of the successful lines of *BT* during its run with the *IUT+*. Each observation event is considered to include its relevant data, such as the PDU types and values, variable values or timer readings involved. As mentioned in 4.1.6, it is sufficient to consider behavior lines which are either pure send, receive, otherwise, timer start, timer cancellation, timeout, readtimer, qualifier or assignment lines. Moreover, we may safely assume that verdicts are assigned exactly in the leaves of the behavior trees. Let *Obs(IUT+,BT)* be the set of all possible observations of the *IUT+* with *BT*.

We assume that *Spec+* is given in a way that permits definition for every (compatible) behavior tree *BT* of the set *Obs(Spec+,BT)* of possible observations of *Spec+* and *BT* in virtual executions of *BT* with the *IUT+*. These are the *valid observations*.

With these basic notions, we can deal with the questions listed above.

WHEN IS A TEST EVENT VALID?

Each valid observation in *Obs(IUT+,BT)* consists entirely of *valid test events*. An *invalid test event* is the first event in the observation which shows that the latter is invalid, i.e. not in *Obs(Spec+,BT)*. In that case, it is unnecessary to assign a validity status to subsequent events in the observation.

WHAT IS A CONFORMANCE REQUIREMENT?

A *conformance requirement* of *Spec* is a (true) property of *Spec*. As the IUT can in part only be accessed through the underlying provider, *testable* conformance requirements are properties of *Spec+*, or the testable part of a property of *Spec* is the resulting set of properties of *Spec+*. Such a property can have many forms. It can be a statement about the structure of an operational model given as a part of

Spec, it can be a mapping from behavior trees *BT* to properties of *Obs*(*Spec*+, *BT*), etc. Very useful are conformance requirements of the form "For all behavior trees *BT* of a given set *BT_Set*, each observation in *Obs*(*IUT*+,*BT*) has the property *P*." Such requirements are closely related to the usual strategy of test cases to try to drive the IUT into particular situations (achieved – with all conforming IUTs – by all behavior trees in *BT_Set*) and to check the validity of the IUT's reactions (which is supposed to fulfill *P*). Let *CR*(*Spec*) be the set of all conformance requirements of *Spec* in a given logical language for such properties.

WHAT IS A TEST PURPOSE?

A *test purpose* is a possible property of a specification, true or false for *Spec*. Many, but not all experts consider test purposes to be nothing but selected conformance requirements, i.e. true properties, and thus elements of *CR*(*Spec*). Of course, as test verdict assignment involves the notion of test purposes, it is important to define unambiguously what a valid test purpose is. At present, it is very much an open question what a test purpose may require and how it is stated. Some of the questions still remaining to be answered in this context are the following:

- Protocol specifications often use an extended finite state machine driven by ASPs and the PDUs contained therein. Should test purposes require driving the "protocol machine" into particular states, or through particular state transitions?
- Should test purposes require performance of certain global event sequences? Should they prescribe the tester's send actions in this sequence, or the complete sequence of tester sends and receives, or, instead, the sends and receives at the IUT's upper and lower boundaries?
- Can the IUT be obliged by the test purpose to choose one of several dynamically valid alternatives? This refers to choices between alternative reactions that, according to the specification and even after the ICS and IXIT are completed, are dynamically left open for the IUT to decide.
- Can test purposes even prescribe event sequences with invalid IUT behavior? This idea was not invented by the authors.

If the third item is answered positively, then test purposes are not necessarily conformance requirements, but at least they must not contradict *CR*(*Spec*). This is a position that can be met frequently. We think it depends on the nature of the alternatives whether or not the IUT can be obliged to make a particular choice. Some alternatives are actually emergency reactions to exceptional resource shortages that cannot be brought about easily in testing. Such alternatives should not be requested by a test purpose. They must be provided for in the test cases, but normally they remain untested, unless they occur by accident. Other choices seem to be made at the discretion of the IUT, and it seems that sometimes the IUT could and should be told which alternative to choose. Perhaps a language element similar to the implicit send should be introduced for this purpose, accompanied by IXIT questions to determine which of these prescriptions the IUT is prepared to follow. Of course, the choices that the IUT can be asked to make in tests must be clearly identified in the OSI specification, which is presently not the case.

At the present stage of standardization, such questions about test purposes may be answered arbitrarily by the test purpose writers, depending on their personal tastes. These questions are made even more important by the fact that many experts silently replace the demand of adherence to conformance requirements, which are well-defined clauses on the OSI specification, by "fulfilment of the test purpose."

WHICH CONFORMANCE REQUIREMENTS DOES A TEST PURPOSE FOCUS ON?

We abbreviate the "conformance requirement(s) on which the test purpose(s) of the test case is (are) focused" as *TPCR(test)*, where

$TPCR(test) \subseteq CR(Spec)$.

This would allow us to sidestep the question above, working instead with the conformance requirements on which the test case is focused. However, as the standard is currently written, this term would be just as undefined. If the test verdict definitions are to be maintained, the "focusing" questions have to be answered.

In the following, we will use the conjunction of all the elements of *TPCR*(*test*), i.e. we will speak of a single property *TPCR*(*test*) instead of a set , such that

$TPCR(test) \in CR(Spec)$.

WHEN IS EVIDENCE OF CONFORMANCE TO A CONFORMANCE REQUIREMENT GIVEN?

If we call the behavior tree defined by the test *BT*(*test*), then the observed test outcome *ObservedOutcome*(*IUT*,*test*) is an element of *Obs*(*IUT+*, *BT*(*test*)). Two intuitively reasonable answers to the above question are:

(a) What was observed is a possible outcome of *test* for at least one possible IUT conforming to *TPCR*(*test*), i.e. formally, for at least one system *Sys*,

$ObservedOutcome(IUT,test) \in Obs(Sys+,test)$, and $TPCR(test)(Sys)$ is true.

(b) What was observed logically implies that the IUT behaved in conformance to the *TPCR*(*test*), i.e. formally, for arbitrary systems *Sys*,

$ObservedOutcome(IUT,test) \in Obs(Sys+,test) \Rightarrow TPCR(test)(Sys)$ is true.

In fact, we would expect (a) to agree with

$ObservedOutcome(IUT,test) \in Obs(Spec+,test)$,

but we omit the theory behind this idea.

WHEN DOES A TEST OUTCOME DEMONSTRATE NONCONFORMANCE WITH RESPECT TO THE TPCR(TEST)?

Here, similarly to the above, the most probable possibilities are

(α) What was observed is a possible outcome of *test* for at least one possible system nonconforming to *TPCR(test)*, i.e. formally, for some *Sys*,

$ObservedOutcome(IUT,test) \in Obs(Sys+,test)$, and NOT $TPCR(test)(Sys)$.

(β) What was observed logically implies that the IUT did not behave in conformance to the *TPCR(test)*, i.e. formally, for arbitrary systems *Sys*,

$ObservedOutcome(IUT,test) \in Obs(Sys+,test) \Rightarrow$ NOT $TPCR(test)(Sys)$.

Again, similarly to the above, we would expect (β) to agree with

$ObservedOutcome(IUT,test) \notin Obs(Spec+,test)$.

These possibilities leave us with four possible (partial) interpretations of the verdict definition. Obviously, it should be clarified in the standard which of them is really meant. But let us try here to assess their relative merits. We will be able to identify three disadvantages of the various interpretations:

(i) The first condition of pass is redundant, because it is implied by the second. This would cast some doubt on the interpretation considered, because it is probable that the standard makers meant something else.

(ii) In some cases, both pass and fail apply, contrary to their characterization as mutually excluding alternatives. This would be counterintuitive, and useless for the conformance testing framework described in Chapter 2.

(iii) Inconclusive never applies, because either pass or fail applies, or both do. This seems definitely not to be intended by the standard, which deals in several places with the verdict inconclusive.

A short analysis shows the following score for the four interpretations:

Interpretation:	(a) + (α)	(a) + (β)	(b) + (α)	(b) + (β)
Disadvantages:	(i), (ii), (iii)	(i), (iii)	(iii)	—

Under the last interpretation,(b) + (β), none of the listed disadvantages applies. It is therefore the only interpretation the authors can recommend, pending further clarification in the standard. With this choice, inconclusive means that, on the basis of the outcome, the IUT may be *conforming to Spec*, i.e. for all possible tests, *Obs(IUT+,test)* $\subseteq$ *Obs(Spec+,test)*, but that the outcome proves neither that it behaved in conformance with *TPCR(test)*, nor that it violated *TPCR(test)* – which is typical if the underlying provider disconnects, and thus messages are lost that might have been invalid or valid.

OTHER POSSIBILITIES FOR VERDICTS

It might even make sense to replace the present (partial) verdict philosophy by a more specific "multiple verdict" approach which answers the following questions:

1. Did the IUT exhibit only valid behaviour *at its boundaries*?
   (verdict suggestion: "VALIDITY: PASS/FAIL/INCONC")

2. Did the IUT fulfil the test purpose?
   verdict suggestion: "PURPOSE: PASS/ FAIL/ INCONC")

Furthermore, in case of INCONCs it would be interesting to know:

3. Was the inability to decide due to exceptional provider behaviour, or is it principally impossible to decide (e.g. if the medium is relatively "opaque," for example, if we apply TTCN to some didactic example outside of OSI, or if the test purpose was ill-chosen).

4.7 Global aspects of test suite execution

In this section, we will make some final remarks about global aspects of test suite execution.

During the entire execution of the test suite, for each successful behavior line,

- a time stamp and, as far as applicable,
- the line number and label,
- the PCO identifier and the data object sent or received,
- the timer causing the timeout,
- assignments and timer operations, and
- the preliminarily or finally assigned test verdict

are recorded in the conformance log. So are all detected test case errors.

We noted previously that test cases are not necessarily executed in the order in which they appear in the ATS document. Information about completed test cases can be saved in test suite variables. The contents of test suite variables survive individual test cases. On the other hand, at the beginning of each test case,

- all test case variables are reset to their initial values, if any are declared,
- all timers are reset to inactive, i.e. they are implicitly cancelled, which includes discarding possibly remaining timeout notifications, and
- all PCO and CR queues are emptied.

The last measure seems reasonable, even though it cannot always guarantee the isolation of test cases, because even the purging of queues between the test cases does not necessarily completely empty the underlying service provider or prevent a faulty IUT from sending more messages. Thus, late events which occur in the wake of the previous test case may still arrive unexpectedly. This forces the test suite writer to deal with these late comers at the beginning of the next test case. It is often impossible to avoid having the last error committed by the IUT in the previous test case look, instead, like the first error of the current test case.

The question of what happens after a test case ends in a FAIL verdict is not quite trivial. Let us look at the moment when the tester discovers invalid behavior of the IUT. A very clean, but also very uneconomical reaction would be to stop testing and send the client home to repair its IUT, which is, of course, not the reaction 9646 prescribes. Why would this be a clean solution? As a matter of fact, after discovering a fault of the IUT, the tester does not have the slightest idea in which state the IUT is. The IUT may have any number of states not provided by its specification and may, from this moment on, behave completely erratically. If the IUT is faulty, there is no guarantee that the tester will ever succeed in bringing it back to a state that remotely resembles one of the valid states! On the other hand, there is some amount of hope, justified by experience, that some drastic measures, like resets and disconnections, will often bring the IUT back to the right behavior, if at least for some time. The whole idea of OTHERWISEs, defaults and postludes hinges on this hope. However if the IUT then immediately fails the next few test cases, too, it makes sense to stop testing and save some of the client's money and the test laboratory's time. Negotiated exits to this effect may be agreed upon in the IXIT.

One of several theoretical justifications for the intuitive "drastic measure" approach mentioned above is the following: it is assumed that the IUT has only the specified number of states, but that it has, besides the specified valid state transitions, a few unspecified state transitions. After discovering an error, the tester may apply a sequence of state transitions that leads from an unknown valid state to a known valid state, provided such a procedure exists.

4.8 Questions of TTCN style

In this section, we recall some of the advice from earlier passages and add a few suggestions that may help to improve the quality and readability of the test suite. They serve to avoid errors in the test suite and to facilitate both its maintenance and its use in profile test specifications. Many more very useful recommendations can be found in [EWOS 93].

Whenever TTCN in the MP form is written or printed on paper, it is most probably because human readers are expected to look at it. Appropriate indentation, maybe even the use of different fonts or font styles, can be of considerable help for these readers.

Naming conventions are important for large documents containing many names and being worked upon by many people. This is typically the case for ATSs. It is recommended that these conventions be specified explicitly, because this facilitates reading the document and permits persons maintaining the document to comply with the conventions when making changes. The usual aspects of naming apply, including the conflicts between different demands, such as making names both short and self-explanatory. Names or name combinations dealing with the same or related objects in different contexts should correspond visibly, for example a test group reference ABC/XYZ/ and a test case identifier ABC_ XYZ_1 of that group.

Comments serve many purposes. They can help especially

- test suite writers and maintainers, whenever they come back to a part written by someone else, or even by themselves, a long time ago,
- ETS realizers, if these comments point out problems to be expected in the derivation process,
- test laboratories, if these comments reference a clause in a standard which justifies test case details that may not be evident for test clients.

9646 even suggests using diagrams in comments, but this may cause problems in the TTCN.MP form.

We remarked in 4.1.3 that testers have to be prepared to receive invalid events. It is not very probable that a tester will ever receive an inopportune type of ASP, e.g. a DATAindication when only a CONNECTindication is valid. As a lower tester, sitting on top of the underlying service provider, this is ruled out by the general assumption that the latter is working correctly, which includes adherence to the service description. For an upper tester, it is, however, conceivable that the IUT does not even conform to the service description. An OTHERWISE covers both invalid ASP types and ASPs of valid type containing an invalid

PDU. Apparently, if only PDUs are used as events, ASP errors committed by the IUT cannot be discovered.

On the other hand, as mentioned in Chapter 2, test purposes may include the verification of correct reaction of the IUT to invalid behavior of the tester. Lower testers can send invalid PDUs. They cannot send invalid ASP types, because the underlying service provider, assumed to work without errors, would not participate. Upper testers should not try to send invalid ASP types: the correct reaction for the IUT would be to refuse participation, and this cannot be expressed in TTCN. Reactions to attempts of ASPs of an invalid type are "local matter". They are specified neither in a protocol specification nor in a service description. From the requirements point of view, they are "invisible" and cannot be an object of investigation in conformance testing.

It seems preferable that only one method of type declaration, tabular or ASN.1 be used for as many declarations as possible, and that the other method only be used where absolutely necessary. Arbitrary mixing of both methods will make constraints hard to read. Type definitions, especially their restrictions and their structure, should correspond as closely as possible to those in the OSI specifications. In particular, restrictions introduced in the OSI specifications should not be deferred to the constraint definition, let alone omitted.

Procedural test suite operation definitions help avoid ambiguities that easily arise in informally defined operations. However, they, too, need careful checking to determine whether or not they are welldefined for each combination of arguments. If they are not, it may help to reduce the ranges of the arguments' types appropriately.

It may be useful to declare more values as test suite parameters than immediately obvious from the OSI specification. Test suite parameters should comprise every value that may not be freely selectable in a test environment, or in an existing or future profile. Premature assignment of specific values may impede the use of the ATS.

The use of test suite constants with self-explanatory names helps to avoid mistyping frequently used explicit values; the values are written only in a single place: the constant definition.

We recommend associating only one behavior, or at least only closely related behaviors, with the same test component in CREATE lines.

For reasons of clarity, the modification mechanism should not be overused in the constraints part. In particular, base constraints should generally supply the majority of the component values of the modified constraints, component values should not be modified twice in the same derivation path, and derivation paths should be short.

Let us remind here again of the rule that constraints should have the same structure as the types they are declared to have. Experience tells that "flat" insertion of a substructure by the macro mechanism and its "structured" insertion as a named component are easily mixed up.

In the dynamic part, test behaviors should be written in such a way as to minimize nondeterminism and to maximize the repeatability of test outcomes. Some binding rules to this effect were introduced in TTCN, as in 4.1.5, where use of send events which might make the test outcome depend on accidental timing conditions is forbidden.

Structuring test cases visibly into preamble, test body, and postamble is recommended. The test body should lead to a test verdict – at least to a preliminary one. Preliminary results should not be used unnecessarily, i.e. in places where they could be omitted without changing the ultimate test verdict. Verdicts in defaults should be assigned particularly carefully, because the default may have been entered from many different places in the test case.

Although recursive tree attachment was used to explain the meaning of constructs, repeats and gotos are often more readable.

Any list of alternatives should be, as far as possible, independent of its order – apart from final OTHERWISEs and TIMEOUTs, of course. Unreachable lines should be avoided, because they merely clutter up the behavior tables. The same holds for superfluous tables in the declarations part, for example definitions of types not used in declarations, or declarations of constraints not used (directly or modified) in the dynamic part.

Deadlocks should be avoided by using OTHERWISE and TIMEOUT appropriately.

The number of necessary test steps can be lowered by skilful parameterization of the test steps and their constraints, making the test steps usable in many places. This, too, should be done with a sense of proportion: it does not pay to throw together different pieces of TTCN serving unrelated purposes, and to select the right behavior by a formal parameter. It does not pay either to create a test step that is used in only a single test case; in that case, a local tree is more appropriate.

Whenever test steps and defaults change test suite or test case variables, it should be pointed out clearly in the comments, maybe even by naming conventions.

Test suite writers have to find the proper balance between the use of local - trees, test steps and defaults. There are many different but equivalent ways of structuring test cases. It is anything but easy to develop a clear style such that readers easily recognize the philosophy behind it. At any rate, make test steps out of local trees if they are identical (possibly up to a few parameters) and appear in several test cases. If there is a unique "final treatment" applicable everywhere in the test body, make it a postamble, rather than a default. Default behavior should not cover the expected valid behavior of the IUT, because this impairs readability of the test case.

As long as the verdict philosophy and the PCO structure are subject to contrary interpretations (cf. 4.6.3 and 4.9), we recommend providing a supplementary scction or document with each ATS, explaining the meaning of verdicts and PCOs adopted in the test suite.

4.9 PCO Problems

The structure of the PCOs, as points of interaction between the tester on one side and the IUT or the underlying provider on the other side, has a marked influence on the formulation of the test cases and on the verdicts assigned to event sequences in a test, cf. e.g.. [BAER 94, BAUM 88, GOTZ 92, PHAL 93, TRET 92]. The 9646 specification of points of control and observation is incomplete. It causes undesirable effects, and so do more specific interpretations of the PCO model. Many of these problems arise not purely within TTCN, but rather from the interplay of 9646 and other OSI standards. Some of them were already addressed in the standardization work, but could not be solved due to reasons of schedule and compromise. These unresolved issues pose considerable problems to test suite writers.

In order to provide a clear and simple picture of these problems, we choose a simple paradigm: we confine ourselves to protocols and do not talk about profiles or information objects etc.; we confine ourselves to single-party testing, in particular to the distributed test method, to layers without substructure, to single-layer and single-protocol IUTs, and to ASPs (and not PDUs) as the primary kind of events dealt with at PCOs. As the tester will be accessing both the IUT and the underlying provider, we will write "IUT/UP" instead of "IUT and underlying provider."

4.9.1 Requirements from outside of 9646 and the interface problem

Let us first recall some principles that can be derived from the OSI Basic Reference Model (BRM) [ISO 7498] and Service Conventions [ISO 10731]:

- OSI does not prescribe actual implementation structures, either of IUTs, or of their interfaces.
- In protocol specifications, ASPs are common atomic actions of a user and a provider at a service boundary. Incomplete ASP attempts are not modelled. ASPs are fictitious constructs, introduced only to define valid open system

behavior operationally. Their realization in actual interfaces is not standardized.

- An OSI service definition constrains the behaviour of the OSI service user by permitting only selected sequences of ASPs.

On the other hand, many OSI implementations provide an upper interface with one-to-one realizations of upper ASPs, occurring in observable orders, which are, for the implementation to be conforming, orders permitted by the service definition, i.e. words of the ASP language. User and provider at the SAP always observe the same order.

In contrast with [ISO 10731], TTCN assumes ASPs to be unidirectional and of two general kinds: Send ASPs are initiated by the tester and transfer information exclusively from the tester to the IUT/UP. Received ASPs are initiated by the IUT/UP and transfer information exclusively from the IUT/ UP to the tester. The rare OSI ASPs that transfer information both ways (for example, the S-SYNCH-MINOR request of the Session service is issued by the user and involves a serial number given to the user) have to be substituted by short sequences of Send and Receive ASPs.

Note that a real tester, or at least the UT component of it, is interacting with the upper service boundary of the IUT. In the OSI BRM, this boundary is only a conceptual abstraction. In the conformance testing configuration outlined at the beginning of this section, it has to be actually realized as an interface. This creates a conflict:

- Outside of 9646, OSI standards explicitly renounce prescribing details of the realization of service interfaces, just as they renounce prescribing details of protocol entity realization.

  Customers acquiring an OSI implementation will ordinarily accommodate any user programs they want to run on top of it to the available interface. They can be seen as "partners" of the implementor.

- But if this boundary has to be accessed by the tester, and the tester's behavior towards the IUT/UP and its possible interactions with them have to be specified precisely, then this indifference is not acceptable. Imagine, to use an analogy, a class of students required to write an exam but being permitted to decide arbitrarily in which language, on which medium and at which speed they will do so. In CTMF, a tester, as an examining instance, would therefore be served best by a complete specification of the upper IUT/UP interfaces both at the UT and at the LT, but this is not provided for in 9646.

  Test laboratories accessing the IUT's upper interface are rather to be seen as "opponents" of the implementor, because it is their task to spot errors. If they cannot get along well with the upper interface, they might consider some of the less successful interactions as IUT errors. Communication between opponents should be regulated carefully, or it will go wrong, and each of the two will blame this on the other one. The LT interface is less of a problem, because we have again the "partner relation" in the tester's local system.

One aspect slightly defuses the "opponent relation" at the UT: Implementors of tester software also want to sell their products and will try hard to enable them to cooperate smoothly at least with the more common types of IUT upper interfaces. But neither will this do justice to a good but unconventional IUT, nor does the appeal to common sense and good will solve all interoperability problems – why else would we need standards and testing, in the first place?

4.9.2 Output queue problems

Similarly to the above conflict between OSI rules and the needs of real testers, there is a conflict between the same rules and the 9646 description of a PCO. TTCN calls for a tester output queue, for ASPs issued from above, situated inside of the IUT/UP, at its upper boundary. This contradicts the OSI rule not to restrict implementations in any way but with respect to their behavior. Which of these two positions has precedence? May we actually ask for a tester output queue in the IUT/UP? And if not: should it be dropped or placed somewhere else?

Moreover, there are other problems about the queue assumption. If the PCO queues are retained, what can be assumed about their capacities? An analysis of the effects of the tester output queue on the observable behaviour of the IUT/ UP depends partly on the capacity of the queue. If it is unbounded, then the only observable effect is that send events from above always succeed "immediately." The guaranteed success of send events is an explicit assumption in TTCN, and, therefore apparently an important intent. Unfortunately, unbounded queues are usually very hard to implement.

If the unbounded tester output queue at the PCO would have to be dropped, we would face a flow control problem. Regular success of tester send events could then only be guaranteed by additional assumptions. However, if we drop the demand that sends always succeed, then some interesting perspectives open up. We will deal with these below.

THE DISCONNECT PROBLEM

In many OSI protocols, as soon as a provider and a user entity have agreed that a connection or association is set up, the provider may issue a disconnect or similar primitive any time it desires. After that, the connection is discontinued, and the user cannot perform any more service primitives with the provider, except those required for establishing a new connection.

Provider disconnects are perfectly valid and do not pose any serious problems, unless seen in conjunction with the TTCN PCO model. If the user is the tester, this means that whenever it tries to send in a test case, the provider may meanwhile have issued a disconnect that is now sitting in the input queue. The send cannot possibly succeed if the output queue is inside the provider, contrary to the required success of the send. A theoretical way out of this dilemma, never to set up a connection fully, is of course not acceptable in real testing.

If the output queue is actually in the IUT/UP, the disconnect problem shows that sends cannot always be successful. Even if the tester has made sure there is no disconnect in the input queue and then turns to the output queue to "send" an ASP, the disconnect may have occurred while the tester "was turning," because disconnects are usually unconfirmed and considered to be possible at any time.

Note that present TTCN even prevents the tester from checking for the absence of disconnects before every send: looking can only happen in a receive line, and receive lines should not be followed by send lines in a list of alternatives, as seen in Subsection 4.1.5.

PCO QUEUES AND SERVICE BOUNDARIES

If we retain the queues, unbounded or not, we are faced with the question of when the ASPs actually "happen" at the service boundary. Do they happen on being put at the back of the input or output queue or on being taken out at its front? In other words: where exactly is the service boundary located at a PCO that is associated with the two queues? We will now investigate merits and disadvantages of various answers to this question.

If we consider unbounded input and output queues, and if we put aside exotic variants (like having the service boundary cut across these queues or permitting unconventional operations on the queues), then the four possibilities shown in Figure 4.9. remain. Each time, the service boundary is represented by the gray line. The "Tester core" in the illustration represents the fictitious entity looking at the queues and executing the behaviour tree, i.e. the left block in Figure 4.2. The whole tester comprises everything above the service boundary, the IUT/UP everything below. For instance, in variant A, the input queue belongs to the tester, and the output queue belongs to the IUT/UP.

If we want to decide which of these PCO structures best fits the requirements posed by various standards as well as the needs of real testing, then we have to check whether each one places the output queue into the IUT/UP, whether it refrains from restricting the implementation beyond the OSI behavior specifications, and whether a send line is always successful. Finally, it should not pose serious practical problems for the real testing of real protocols. Table 4.17 shows the assessment of A-D.

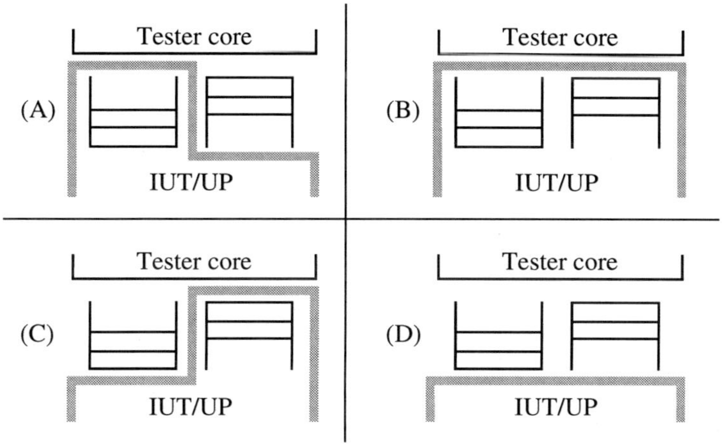

FIGURE 4.9. PCO structures relating two unbounded queues to the service boundary

TABLE 4.17. Some pros and cons of possible PCO structures

Demands to be met \ PCO structure	A	B	C	D
Output queue in the IUT/UP	+	+	–	–
No implementation details of IUT/UP	–	–	–	+
Sends always successful	–	–	+	+
Unconditionally implementable	–	–	–	–
No “IUT/UP Disconnect” problem	–	–	+	+

If we simply count the arguments introduced in favour and to the disadvantage of the four PCO structures, then D gains a narrow victory, with C as runner-up. None of the four candidates meets more than three out of the five demands. More than four cannot be expected, anyway, because they are in conflict, as shown before. We desisted from deriving a linear ranking among the four models.

4.9.3 Alternative PCO structures

ATOMIC ASPS

Let us assume for the moment that, contrary to 9646, we apply the abstract service boundary model to TTCN test specifications. With our simple paradigm, we could equate PCOs with SAPs and ASPs with atomic interactions of the tester and the IUT/UP. Even such a very abstract view leaves room for conflicting interpretations. A real tester can always observe more than which ASPs have completed in which order, i.e. more than is happening abstractly at the service boundary in the BRM. For example, if the IUT is expected to issue an ASP to the UT by a certain time, but erroneously does not, then the tester can use a clock to find out and call the test case failed. But just how much more can it observe? With respect to send events, the following questions arise:

(a) What happens if the user of the IUT, in our configuration the tester, issues a valid ASP at UT, and the IUT (which does not necessarily implement its specification), will not accept it? Can the two deadlock within this ASP? Can the user recognize that the provider declines and call this a failed test case? These questions affect the usability of test cases as well as verdict assignment in real testing, and therefore cannot be put aside as mere implementation details.

(b) What if the IUT, e.g. in some kind of backpressure situation, cannot accept the ASP right now, but would in principle accept it later, and therefore defers acceptance without downright declining it? Will the user recognize the delay? If so, will it wait arbitrarily long for the ASP to complete or stop the attempt? These questions are of similar rank as those under (a).

Assume test case execution is simply delayed in a send event until the IUT/UP is able and willing to accept it. At least this might be a variant view of TTCN operational semantics. In reality this delay could be filled with local interactions. Of course, if a send can delay arbitrarily, it can also delay for unreasonably long times. Testers, however, always have only a limited amount of time available. Hence, delaying send events calls for a facility that is presently not permitted in TTCN: something like a "send-timeout" or an "elapsed-patience-default," which is attempted as an alternative to unsuccessful sends, in order to prevent excessive waiting for a send to succeed.

This approach has a subtle disadvantage that could be avoided by a different philosophy: the outcome of an attempt of the current list of alternatives is no longer determined by the snapshot, because the snapshot does not tell whether or when a send is finally successful. The disconnect problem explains why this knowledge cannot be part of a snapshot. Suppose on the other hand, we would insist that the snapshot determines the success of the alternatives. The following approach discusses changes that could be made to guarantee this property.

PCO BUFFER

Some experts intuitively interpret the PCO queues as having at most one element. Even though this view is not backed by the standard, it does have merits of its own.

If the output queues actually have a capacity of one – they could then be called output buffers – then the snapshot could include these buffers, or at least the information pertaining to whether they are full or empty. The tester could go through any list of alternatives and decide about their success on the basis of the snapshot. Sends would not always succeed, specifically if the output buffer at the respective PCO is not reported as empty in the snapshot (although it might have become empty after the snapshot was taken, which is in keeping with snapshot semantics). The tester would have to try to get the contents of its output buffers across the service boundary, towards the IUT/UP, continually, preferably concurrently with its other activities.

If input queues were also bounded by a capacity of one ASP, this would require that the IUT/UP also not send at an arbitrary speed. It should be possible to write the test case in such a way that a conforming IUT would not issue too many ASPs too fast. In practice, there will be provisions to prevent that anyway, because IUTs which insist that flooding its user entities with ASPs at uncontrollable rates would have problems on the market.

SPECIFIC INTERFACE DEFINITIONS

Theoretically, one could implement any number of different types of upper interfaces for OSI implementations, making it hard for real testers both to connect to the SAPs at all, and, sometimes, even to find out whether they have found an error or only misunderstood the interface. Prior interface testing, not standardized in any way, offers only ad hoc relief. Standardized interfaces, no matter whether uniform throughout OSI or protocol-specific, would definitely help. So would, alternatively, a standardized *interface definition language*, permitting IUTs to be accompanied by a sufficient description of the realization of the upper service boundary, as a part of the ICS, for example.

The lack of such commitments was probably the reason that 9646 came up with its (somewhat sketchy) structure of PCOs, giving service boundaries slightly more structure than that provided by the atomic interactions.

Our discussion on the structure of PCOs showed that the requirements of 9646 and those of the BRM and service conventions are conflicting and not fully reconcilable. Something has to give, and we have tried to show some scenarios of what could be achieved if various requirements were relaxed or, vice versa, if interface structures were described more rigorously.

Subsections 4.6.3 and Section 4.9 are based on the paper [BAUM 94].

CHAPTER 5. TEST DOCUMENTS FOR AN EXAMPLE PROTOCOL

In this chapter, the OSI conformance testing methodology and especially the test notation TTCN are applied to a simple example protocol, called *EXPRO*. This protocol is considerably less complex, both in protocol functionality and in data structures, than real OSI protocols. First we present the protocol and service specification of EXPRO. Then we show in detail what test related documents, such as TSS&TP, PICS, PIXIT, and an ATS, may look like.

Readers who want to test their understanding of the material presented in Chapters 2 to 4, are recommended to proceed as follows:

- Read the specifications in Section 5.1.
- For each of the test documents, i.e. for TSS&TP, PICS, PIXIT and ATS, one after the other, first write your own version, based on the previous documents, and then, before you go on to the next document, compare your version with the one suggested in the book. Use the distributed test method in the ATS. Assume an initiator entity is tested (cf. 5.1).

Readers may ask why we did not use a protocol which is used in practice as an example. The reason is that OSI test suites are normally very bulky documents, often encompassing several hundred pages, and are therefore not suitable for even a partial presentation and discussion within a tutorial text. Therefore, we decided to specify a reasonably small example protocol, just complex enough to allow readers to practice the most important mechanisms of test specification. Still, for reasons of volume, some passages of the test documents have been abridged or completely omitted.

Many useful examples of protocols and services were invented for didactic purposes. Inres [HOGR 91], CAB [BGP 93] and others [TURN 93] were introduced as simple examples with some OSI-typical features, in order to serve as a basis for formal specification or test generation case studies. Test cases for the CAB protocol were derived by a test generation tool from formal descriptions in the form of higher level Petri nets. The sole protocol task of EXPRO is to provide a primitive form of global flow control.

5.1 The EXPRO protocol

The overall architecture of EXPRO is shown in Figure 5.1. *EXPRO initiator* and *responder* protocol entities make use of the service provided by an underlying layer, called *medium*, and provide the *EXPRO service* to their *EXPRO initiator* and *responder users*. All types of service primitives, which are exchanged "vertically," as well as all types of PDUs, which are exchanged "horizontally," are listed in the figure (SPs with abbreviated names).

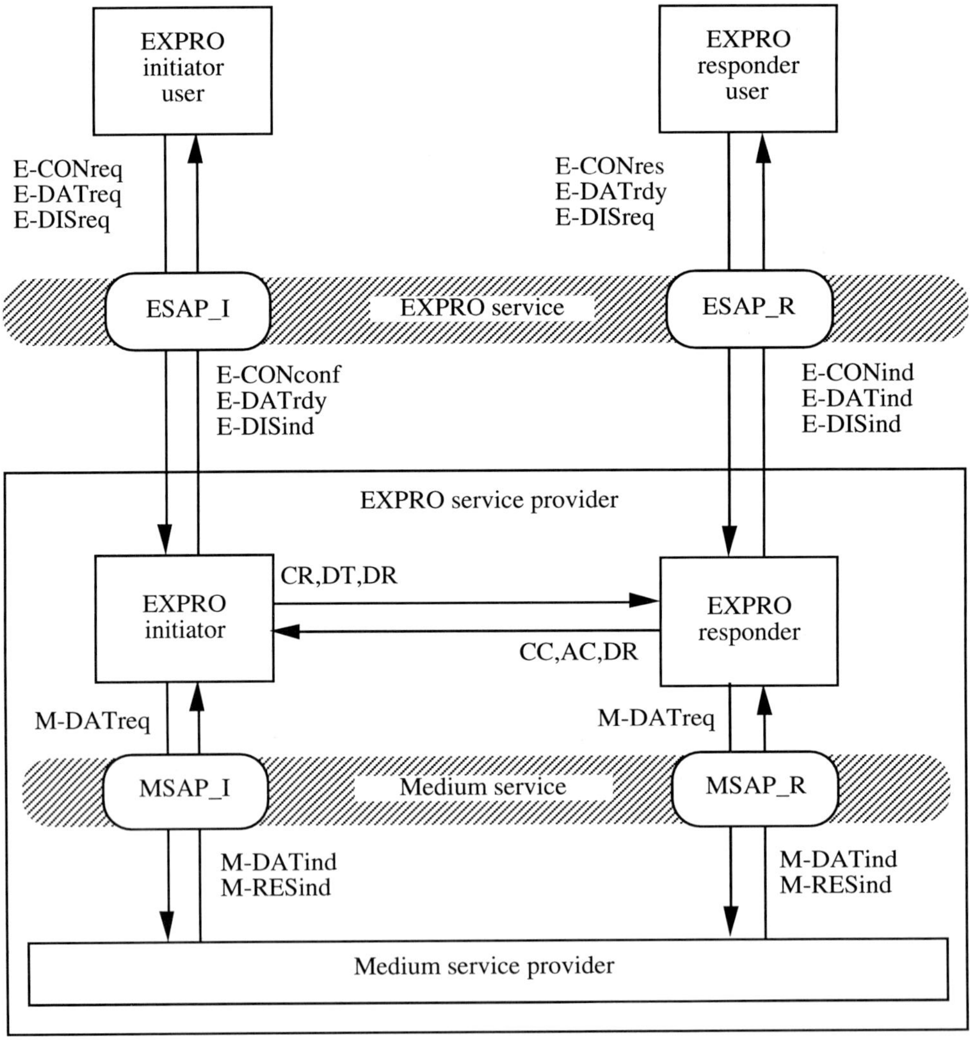

FIGURE 5.1. Basic architecture of the EXPRO system

The specification of EXPRO consists of three parts:

- The service provided by the EXPRO layer is specified in document *EServ*.
- The medium service is specified in document *MServ*.
- The functions of the EXPRO layer protocol are specified in document *EProt*.

5.1.1 *Definition of the EXPRO service*

DOCUMENT ESERV

ESERV.1 ABBREVIATIONS AND DEFINITIONS

The following definitions and abbreviations apply:

EC	EXPRO connection
ES	EXPRO service
ESAP_I	EXPRO service access point at the initiator
ESAP_R	EXPRO service access point at the responder
ESDU	EXPRO service data unit
ESDUM	maximum ESDU length in octets = 255
EIU	EXPRO initiator user, the ES user that initiates an EC establishment request
ERU	EXPRO responder user, the ES user with which an EIU wishes to establish an EC
I_ADDR	calling address, the address of an EIU, identifying a connection end point of ESAP_I
R_ADDR	called address, the address of an ERU, identifying a connection end point of ESAP_R
ORI	originator parameter
REA	reason parameter

ESERV.2 OVERVIEW OF THE EXPRO SERVICE

The EXPRO service includes the service primitives and their parameters listed in Table 5.1, and offers the following features to an ES user:

- establishment of an EC with another ES user for the purpose of sending the latter ESDUs (multiple ECs between the same ES users are not supported),

- transfer of ESDUs from an EIU to an ERU, without duplication or reordering, but with possible loss of data,
- unconfirmed termination of an EC by the EIU, the ES provider, or both.

TABLE 5.1. EXPRO service primitives

Primitives		**Parameters**	**Types**
E-CONNECT	request	I_ADDR	OCTETSTRING[1]
		R_ADDR	OCTETSTRING[1]
	indication	I_ADDR	OCTETSTRING[1]
		R_ADDR	OCTETSTRING[1]
	response	I_ADDR	OCTETSTRING[1]
		R_ADDR	OCTETSTRING[1]
	confirm	I_ADDR	OCTETSTRING[1]
		R_ADDR	OCTETSTRING[1]
E-DATA	request	I_ADDR	OCTETSTRING[1]
		R_ADDR	OCTETSTRING[1]
		ESDU	OCTETSTRING[0..ESDUM]
	indication	I_ADDR	OCTETSTRING[1]
		R_ADDR	OCTETSTRING[1]
		ESDU	OCTETSTRING[0..ESDUM]
	ready	I_ADDR	OCTETSTRING[1]
		R_ADDR	OCTETSTRING[1]
E-DISCONNECT	request	I_ADDR	OCTETSTRING[1]
		R_ADDR	OCTETSTRING[1]
		REA	OCTETSTRING[1]
		ORI	OCTETSTRING[1]
	indication	I_ADDR	OCTETSTRING[1]
		R_ADDR	OCTETSTRING[1]
		REA	OCTETSTRING[1]
		ORI	OCTETSTRING[1]

ESERV.3 SEQUENCE OF EXPRO SERVICE PRIMITIVES

This clause defines the constraints on the sequences in which the ES primitives may occur. The order of SPs at the different SAPs is specified in Figure 5.2, employing time sequence diagrams, as defined in [ISO 10731]. The initiator side is on the left. All allowed local sequences of ES primitives occurring at ESAP_I or at ESAP_R are defined, by means of state transition diagrams, in Figure 5.3.

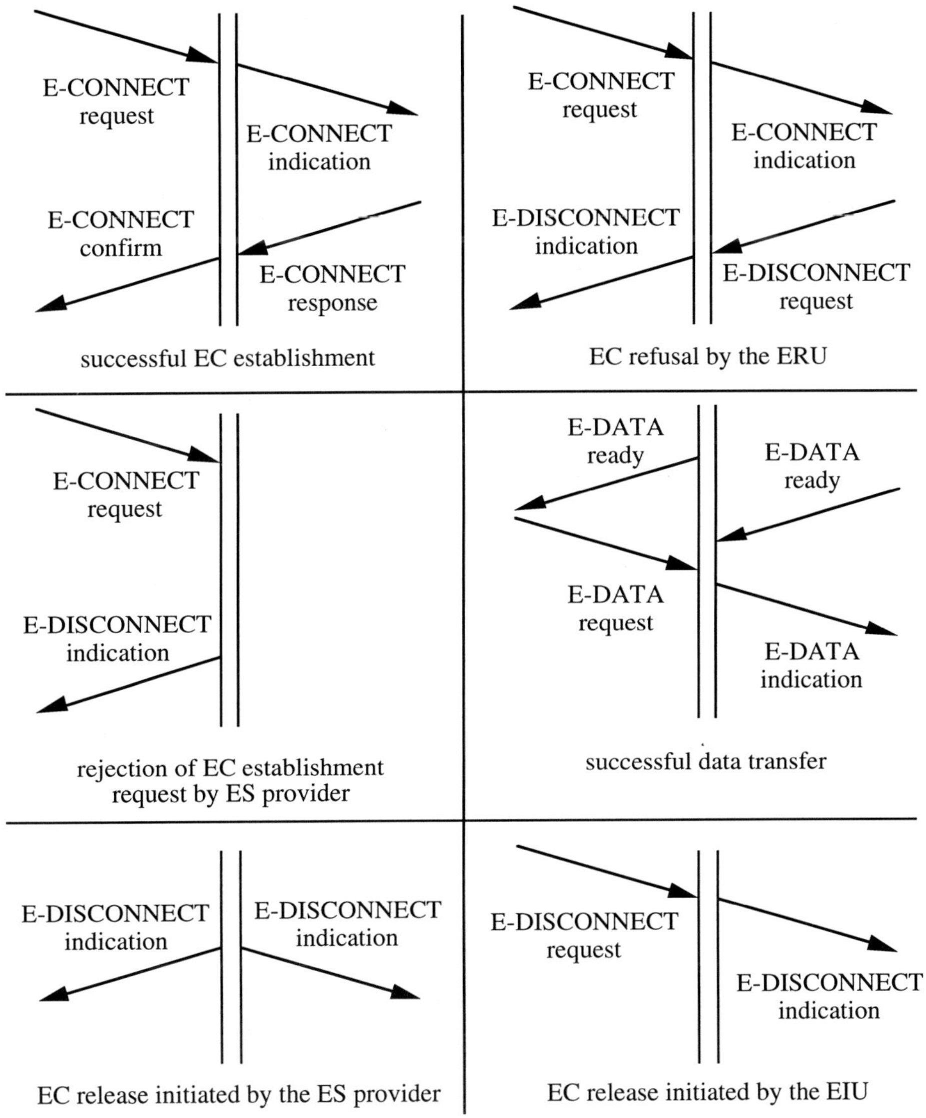

FIGURE 5.2. ES time sequence diagrams

ESERV.4 EXPRO CONNECTION ESTABLISHMENT

The EXPRO service is connection oriented and asymmetric. The ES can be accessed at two SAPs. An EIU that wants to communicate with an ERU via the ES initiates an EC at one SAP (the left one in Fig. 5.1, ESAP_I). At the other SAP (ESAP_R), the ERU can accept or reject the connection. Acceptance is achieved by an E-CONNECTresponse. Rejection is achieved by an E-DISCON-

NECTrequest with the parameters ORI='01'O, for ERU, and REA='01'O, for "connection refusal". As soon as the EIU receives the E-CONNECTconfirmation, the establishment phase is completed successfully.

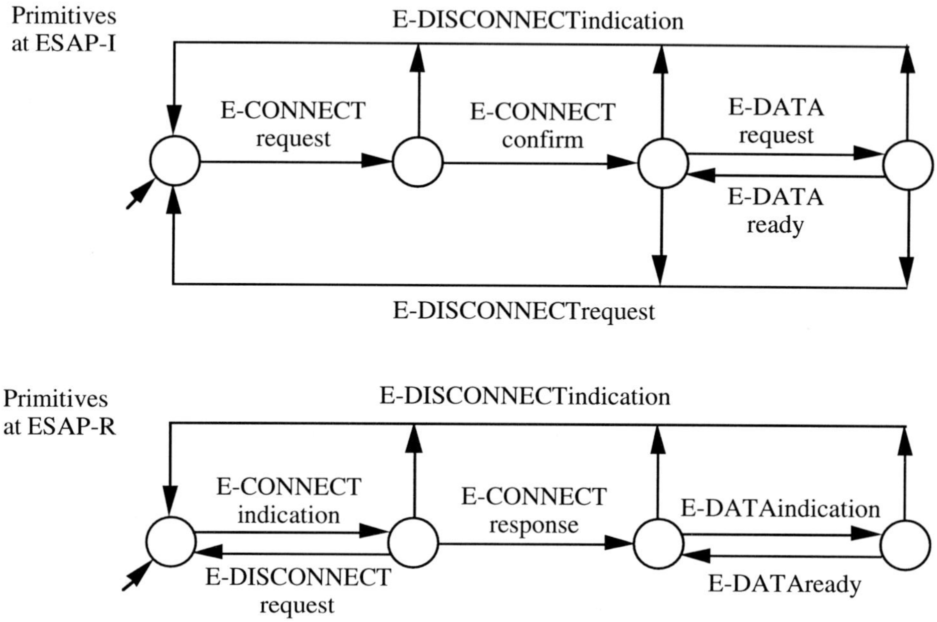

FIGURE 5.3. State transition diagrams for allowed sequences of ES primitives

ESERV.5 DATA TRANSFER PHASE

After a successful completion of the establishment phase, the data phase is entered. In this phase, the EIU is allowed to hand over ESDUs to the initiator as parameters in E-DATArequest primitives. The passing of E-DATArequests is locally flow-controlled by the service primitive E-DATAready. The desired global service during one connection consists in the ultimate delivery of these ESDUs to the ERU as parameters of E-DATAindication primitives. The receipt of E-DATAindication primitives is also locally flow-controlled by the service primitive E-DATAready.

ESERV.6 DISCONNECT PHASE

Normal release of an EC can only be initiated by the EIU in the data phase, by means of an unconfirmed E-DISCONNECTrequest with the parameters ORI = '00'O, for EIU, and REA = '00'O, for "normal release."

Abnormal release of an existing EC can only be initiated by the ES provider by means of an unconfirmed E-DISCONNECTindication with the appropriate values for the originator and reason parameters at both endpoints of the connection. The following values for the originator and reason parameters are distinguished:

REA '00'O (normal), '01'O (connection refusal), '02'O (procedure error), '03'O (medium reset);

ORI '00'O (EIU), '01'O (ERU), '02'O (initiator), '03'O (responder), '04'O (medium).

END OF DOCUMENT ESERV

5.1.2 *Definition of the medium service*

DOCUMENT MSERV

The medium service permanently connects two end systems at the SAPs MSAP_I and MSAP_R. It is symmetrical and operates connectionless, unconfirmed and reliably, up to the M-RESETs.

MSERV.1 ABBREVIATIONS AND DEFINITIONS

MS	medium service
MSAP_I	medium service access point at the initiator
MSAP_R	medium service access point at the responder
MSDU	medium service data unit
MSDUM	maximum MSDU length = 260

MSERV.2 OVERVIEW OF THE MEDIUM SERVICE

The medium service includes the service primitives listed in Table 5.2 and offers to the MS-users:

- data transfer of MSDUs between MSAP_I and MSAP_R without duplication or reordering and with a guaranteed maximum transit delay.
- In exceptional situations, the medium may issue one M-RESET primitive each at both MSAP_I and MSAP_R, which entails loss of all MSDUs under way between these SAPs.

TABLE 5.2. Medium service primitives

Primitives		**Parameters**	**Types**
M-DATA	request	MSDU	OCTETSTRING[0..MSDUM]
	indication	MSDU	OCTETSTRING[0..MSDUM]
M-RESET	indication	REA	OCTETSTRING[1]
		ORI	OCTETSTRING[1]

MSERV.3 SEQUENCE OF MEDIUM SERVICE PRIMITIVES

This clause defines the constraints on the sequences in which the MS primitives may occur. The order of service primitives at either SAP is specified in Figure 5.4, employing time sequence diagrams, as defined in [ISO 10731]. Left and right can be interchanged; mirror images are not shown. The allowed local sequences of MS primitives at each SAP are defined by means of the state transition diagram in Figure 5.5.

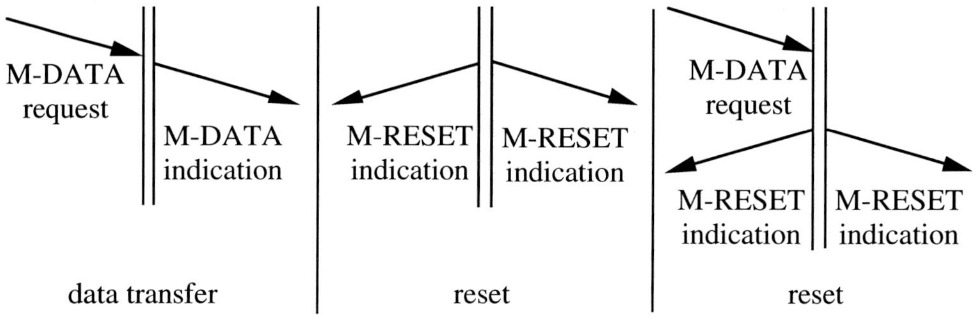

FIGURE 5.4. MS time sequence diagrams

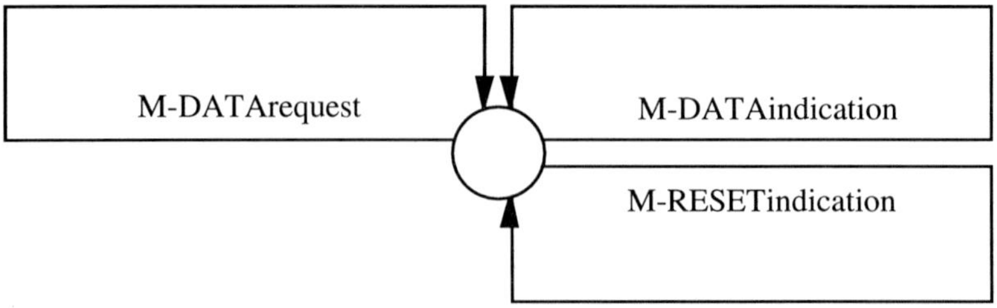

FIGURE 5.5. State transition diagram for allowed sequences of MS primitives

MSERV.4 DATA TRANSFER PHASE

The medium is permanently in the data phase, where both the medium user at MSAP_I and at MSAP_R are allowed to hand over MSDUs consecutively to the medium as parameters of M-DATArequest primitives. The desired global service consists in the delivery of these MSDUs to the other user. The data phase is terminated and restarted by medium resets, with loss of all data under way.

END OF DOCUMENT MSERV

5.1.3 Specification of the EXPRO protocol

DOCUMENT EPROT

The EXPRO protocol supports the EXPRO service defined in EServ. Information is transferred to and from the ES-user in the ES-primitives listed in Table 5.1. The EXPRO protocol makes use of the medium service defined in MServ.

EPROT.1 DEFINITIONS AND ABBREVIATIONS

This document makes use of all terms and abbreviations defined in EServ and MServ, plus the following ones:

AC	acknowledgement EPDU
CC	connection confirm EPDU
CR	connection request EPDU
DR	disconnect request EPDU
DT	data EPDU
EI	EXPRO initiator entity
EPDU	EXPRO protocol data unit
ER	EXPRO responder entity
PID	PDU identifier
RQ	the FIFO queue of the ER for undelivered DTs. It always contains (VR-VU) DTs.
UD	the user data field in a DT
VR	local variable of the ER, used to count the number of DTs received after the last AC
VS	local variable used by the EI, used to count the remaining number of DTs which may be transmitted unacknowledged

VU	local variable of the ER, used to count the number of E-DATA-indications after the last AC
WS	the optional window size field in a CC
WSI	window size granted to initiator. It defines the maximum number of DTs that the EI is allowed to send consecutively to the ER. The sending and reception of further DTs is enabled by ACs transmitted from the ER to the EI. The ER shall transmit an AC as soon as WSI DTs have arrived since the last AC and if the ER is able to receive WSI further DTs. By default, WSI=1. In the CC EPDU, the responder may raise WSI to any value 1≤WS≤255.
WSR	window size implemented in the responder, 1≤WSR≤255. In the CC EPDU, the responder may send WS with WS:=WSR.

EPROT.2 GENERAL PROPERTIES OF THE PROTOCOL

The EXPRO protocol is a connection oriented protocol that operates between two protocol entities, called the EXPRO initiator (EI) and responder (ER). The protocol entities communicate by exchanging the protocol data units CR, CC, DT, AC and DR. The PDUs are listed, together with their fields and their relationships to ES-primitives, in Table 5.3.

The communication between the two protocol entities takes place in three phases: the connection establishment phase, the data transmission phase and the disconnection phase:

- The purpose of the connection establishment phase is to establish an EC from an EIU to an ERU and to provide a means to distinguish EPDUs of different ECs (see EProt.3.1).
- The purpose of the data transfer phase is to permit the transmission of ES-DUs from an EIU to an ERU, connected by the EC. This transmission provides
  - global flow control by means of a simple acknowledgement mechanism, used to regulate the flow of DT EPDUs from the EIU to the ERU,
  - EC identification by address information in every PDU, to identify the unique EC between the EIU and the ERU.
- The purpose of the disconnection phase is to perform either
  - normal release of an EC, i.e. an unconfirmed disconnection of an EC, initiated by the EIU (see EProt.3.2), or
  - abnormal release of an EC, i.e. an unconfirmed disconnection of an EC, initiated by the EI or the ER after the occurrence of a procedure error or after the reception of an M-RESETindication initiated by the medium.

TABLE 5.3. EXPRO PDUs, their fields and related ES-primitives

PDU Name	Function	PDU Field	Related ES Primitives
CR	connection request	PID I_ADDR R_ADDR	E-CONNECTrequest, E-CONNECTindication
CC	connection confirmation	PID I_ADDR R_ADDR	E-CONNECTresponse, E-CONNECTconfirm
DR	disconnect request	PID I_ADDR R_ADDR REA ORI	E-DISCONNECTrequest, E-DISCONNECTindication
DT	data transfer	PID I_ADDR R_ADDR UD	E-DATArequest, E-DATAindication, E-DATAready
AC	acknowledgement	PID I_ADDR R_ADDR	E-DATAready

EPROT.3 ELEMENTS OF PROCEDURE

The state transitions of the initiator and the responder entities are shown in Tables 5.4 and 5.5, where some obvious abbreviations are used for service primitives. The input events are usually written as SPs, but "M-DATind with MSDU = encoding of <PDU name>" is abbreviated as "<PDU name>." In the actions, "<SP name>" means "issue <SP name>," and "<PDU name>" means "issue M-DATreq with MSDU = encoding of <PDU name>," such that PDU names can be used besides SP names in both input events and actions.

Invalid PDUs include inopportune PDUs and PDUs with syntactically correct but semantically invalid field values, for example, if the EI receives in state "wait_CC" a DR with ORI='00'O (i.e. denoting EI).

EPROT.3.1 CONNECTION ESTABLISHMENT PHASE

The following states are used to describe the connection establishment phase at the initiator and responder entities:

idle_I	Initial state of the initiator
wait_CC	The initiator is waiting for a CC from the responder.

data_I The initiator is able to transmit DTs.

idle_R Initial state of the responder

wait_resp The responder is waiting for an E-CONNECTresponse from the responder user.

data_R The responder is able to pass an E-DATAindication to the responder user.

The values of the calling (I_ADDR) and called addresses (R_ADDR) of the E-CONNECTrequest primitive are carried in every PDU. The optional window size field WS in the CC PDU may be used to extend the upper limit of the number of unacknowledged DTs to a number higher than 1.

TABLE 5.4. State transitions of the initiator

State	**Input events and conditions**	**Actions**	**Target State**
idle_I	E-CONreq	CR, WSI:=1	wait_CC
	M-RESind	ignore	idle_I
	M-DATind	ignore	idle_I
wait_CC	CC	E-CONconf, if WS present: WSI:=WS, VS:=WSI	data_I
	DR	E-DISind with the same REA and ORI as in DR	idle_I
	M-RESind	E-DISind with the same REA and ORI as in M-RESind	idle_I
	invalid PDU	E-DISind and DR with REA= '02'O (procedure error) and ORI='02'O (initiator)	idle_I
data_I	E-DATreq $\wedge$ VS>1	DT, VS:=VS-1, E-DATrdy	data_I
	E-DATreq $\wedge$ VS=1	DT, VS:=0	wait_AC
	E-DISreq	DR with the same REA and ORI as in E-DISreq	idle_I
	M-RESind	E-DISind with the same REA and ORI as in M-RESind	idle_I
	DR	E-DISind with the same REA and ORI as in DR	idle_I
	invalid PDU	E-DISind and DR with REA='02'O (procedure error) and ORI='02'O (initiator)	idle_I
wait_AC	AC	E-DATrdy, VS:=WSI	data_I
	E-DISreq	DR with the same REA and ORI as in E-DISreq	idle_I
	M-RESind	E-DISind with the same REA and ORI as in M-RESind	idle_I
	DR	E-DISind with the same REA and ORI as in DR	idle_I
	invalid PDU	E-DISind and DR with REA='02'O (procedure error) and ORI='02'O (initiator)	idle_I

TABLE 5.5. State transitions of the responder

State	Input events and conditions	Actions	Target State
idle_R	CR	E-CONind	wait_resp
	M-RESind	ignore	idle_R
	invalid PDU	ignore	idle_R
wait_resp	E-CONres	CC with WS=WSR (WS may be omitted if WSR=1), VR:=0, VU:=0, empty RQ	data_R
	E-DISreq	DR with the same REA and ORI as in the E-DISreq	idle_R
	M-RESind	E-DISind with the same REA and ORI as in the M-RESind	idle_R
	DR	E-DISind with the same REA and ORI as in the DR	idle_R
	invalid PDU	E-DISind and DR with REA= '02'O (procedure error) and ORI='03'O (responder)	idle_R
data_R	DT $\wedge$ VR<WSR	enqueue DT to RQ, VR:= VR+1	data_R
	VU = WSR	AC, VR:=0, VU:=0	data_R
	VR > VU	E-DATind with dequeued front element of RQ, VU:=VU+1	wait_ready
	M-RESind	E-DISind with the same REA and ORI as in the M-RESind	idle_R
	DR	E-DISind with the same REA and ORI as in the DR	idle_R
	invalid PDU or (DT $\wedge$ VR=WSR)	E-DISind and DR with REA='02'O (procedure error) and ORI='03'O (responder), empty RQ	idle_R
wait_ready	DT $\wedge$ VR<WSR	enqueue DT to RQ, VR:=VR+1	wait_ready
	E-DATrdy		data_R
	M-RESind	E-DISind with the same REA and ORI as in the M-RESind	idle_R
	DR	E-DISind with the same REA and ORI as in the DR	idle_R
	invalid PDU or (DT $\wedge$ VR=WSR)	E-DISind and DR with REA='02'O (procedure error) and ORI='03'O (responder), empty RQ	idle_R

EPROT.3.2 DATA TRANSMISSION AND DISCONNECT PHASE

In addition to the states listed in EPROT 3.1, the following states are used to describe the data and disconnect phase at the initiator and responder:

wait_AC The initiator is waiting for an AC from the responder, in order to continue with the transmission of further DTs.

wait_ready The responder is waiting for the E-DATAready primitive from the ERU.

The local variables VS, VR and VU are used as described in their definitions and in the state transitions tables, in order to assure that the EI never sends more unacknowledged data than permitted and that the ER sends acknowledgements whenever the EI cannot continue without them.

There can be three different initial causes for disconnection:

- an E-DISreq from the EIU: in this case, a DR is sent to the peer entity.
- discovery of a procedure error by EI or ER: in this case, both a DR is sent to the peer entity, and an E_DISind is issued.
- an M-RESind from the Medium issued to both EI and ER: in this case, an E-DISind is issued on each side.

After the reception of a DR from the peer entity, an E_DISind is issued. Any entity initiating disconnection will indicate the appropriate originator and reason parameters or fields. An EI or ER participating in a disconnection initiated at the other SAP or by the medium will maintain the received originator and reason informations in the SPs it issues or the PDUs it sends to complete disconnection. DRs under way in the medium can be lost during a reset, just as any other data.

In the connection establishment phase, there is not yet any connection that could be released. Therefore, a DR sent in this phase by the ER as a negative response to the CR plays the role of a "negative CC" and is not part of the disconnection of an existing connection.

On the EXPRO user level, disconnection leads to the loss of the DTs sent by the EI and not yet delivered as E-DATind to the ERU, because the ER subsequently ignores anything but CR and empties its queue RQ of undelivered data. The only moments the EIU can be sure that all its previously sent ESDUs have arrived at the ERU are when it gets an E-DATrdy after sending a multiple of WSI ESDUs. This may not be completely satisfactory, but it simplifies the EXPRO specification considerably.

EPROT.4 STRUCTURE AND ENCODING OF EPDUS

Information related to the structure and encoding of the EPDUs is given in Tables 5.3 and 5.6. All fields, with the exception of WS, are mandatory. The total number of octets in any EPDU shall not exceed MSDUM. The length of the user data field of a DT can therefore not exceed ESDUM = MSDUM - 5 = 255, because of 5 leading octets for PID, two addresses, the UD field identifier, and the UD field length.

PDU encodings are produced by concatenation of the encodings of their fields and subfields, in the order shown in the table. The PDU fields PID (PDU identifier), I_ADDR (EIU address), and R_ADDR (ERU address) are unstructured data types, whereas the user data (UD) and window size (WS) fields are structured data types, composed of identifier, length (only UD) and contents. For the purpose of demonstration, the structures are chosen to be slightly more complex than absolutely necessary. Thus they permit a greater variety of PDU errors.

TABLE 5.6. Encoding of EPDUs

PDU field	PDU subfield	Encoding information	Comments
PID		OCTETSTRING[1] '00'O	connect request
		OCTETSTRING[1] '01'O	connect confirm
		OCTETSTRING[1] '02'O	disconnect request
		OCTETSTRING[1] '03'O	data
		OCTETSTRING[1] '04'O	acknowledgement
I_ADDR		OCTETSTRING[1]	address '00'O is not allowed
R_ADDR		OCTETSTRING[1]	address '00'O is not allowed
UD			user data field consisting of
	UD_I	OCTETSTRING[1] '00'O	UD field identifier
	UD_L	OCTETSTRING[1]	UD field length<ESDUM
	UD_C	OCTETSTRING[udl]	UD field contents udl=integer value of UD_L
WS			optional window size field consisting of
	WS_I	OCTETSTRING[1] '01'O	WS field identifier
	WS_C	OCTETSTRING[1]	WS field contents [1..255]
REA		OCTETSTRING[1]	reason
ORI		OCTETSTRING[1]	originator

EPROT.5 SYSTEM PARAMETERS

EXPRO makes use of one system parameter, WSR, which is in fact a parameter of the responder entity ER. WSR may have any value between 1 and 255.

EPROT.6 CONFORMANCE

A system claiming to implement the procedures specified in this example document shall comply with the requirements EProt.6.1 to EProt.6.4.

EPROT.6.1 The system shall implement either the initiator or the responder capabilities.

EPROT.6.2 It shall operate the procedures defined in EProt.3, depending on the role of the system.

EPROT.6.3 It shall perform the encoding and decoding of EPDUs as defined in EProt.4.

EPROT.6.4 It shall support the system parameter as defined in EProt.5, depending on the role of the system.

END OF DOCUMENT EPROT

5.2 EXPRO Test suite structure and test purposes

DOCUMENT ETP

ETP.1 INTRODUCTION

This document specifies the test suite structure and the set of test purposes (TSS&TP) as the basis for developing the standardized conformance test suites used to evaluate conformance to EProt.

ETP.2 TEST SUITE STRUCTURE

The EXPRO test suite consists of a number of test groups and test cases, arranged in a hierarchical manner. An outline of the EXPRO test suite structure is shown in Table 5.7.

- Level 1 is divided into two main test groups, associated to initiator and responder test group objectives.
- Level 2 distinguishes between objectives for the connection establishment, disconnection and data transfer phases.
- Level 3 lists test group objectives with respect to IUT behaviour after the receipt of only valid or also invalid PDUs (including syntactically invalid and inopportune PDUs).
- Level 4 concentrates on further details of the procedures, like connection acceptance, connection rejection, transfer of particular data, etc. In the case of the level 3 test group "invalid," level 4 distinguishes between possible error categories, such as PDUs with unknown protocol identifier, missing, duplicated and reordered fields, field contents errors and inopportune PDUs.

The test group references are denoted as 3-character mnemonics and correspond to test group objectives. They are also used in the description of the test purposes. At least one test purpose and its corresponding test case have to be specified for every lowest test group objective (level 4 in our example).

TABLE 5.7. EXPRO test suite structure

Level 1	Level 2	Level 3	Level 4	Test group objective
INI/	test group for initiator behaviour			
	CON/	test group for connection establishment		
		VAL/	the LT is sending only valid PDUs	
			ACC/	acceptance of connection
			REJ/	rejection of connection
		INV/	the LT is sending also invalid PDUs	
			UPI/	unknown PID
			UFI/	unknown field
			MFI/	missing field
			DFI/	duplicated field
			RFI/	reordered fields
			FCE/	field contents error
			INO/	inopportune PDU
	DIS/	test group for disconnection in the data phase		
		VAL/	the LT is sending only valid PDUs	
			IUD/	initiator user disconnect
			RED/	responder disconnect
		INV/	the LT is sending also invalid PDUs	
			UFI/	unknown field
			MFI/	missing field
			DFI/	duplicated field
			RFI/	reordered fields
			FCE/	field contents error
	DAT/	test group for data transfer		
		VAL/	the LT is sending only valid PDUs	
			ACK/	acknowledgement mechanism
		INV/	the LT is sending also invalid PDUs	
			UPI/	unknown PID
			UFI/	unknown field
INI/	DAT/	INV/	MFI/	missing field
			DFI/	duplicated field
			RFI/	reordered fields
			FCE/	field contents error
			INO/	inopportune PDU
RES/	test group for responder behaviour			Exercise for the reader

ETP.3 TEST PURPOSES

In this section, the test group objectives and the test purposes for the EXPRO protocol are specified. The corresponding test group references have been added in parentheses. Test purpose texts are written in italics.

ETP.3.1 Test purposes for initiator IUTs (INI/)

ETP.3.1.1 Connection establishment (INI/CON/)

ETP.3.1.1.1 Response to the receipt of valid PDUs (INI/CON/VAL/)

ETP.3.1.1.1.1 Acceptance of connection (INI/CON/VAL/ACC)

Ensure that the IUT transmits a CR PDU and reacts properly to the receipt of a CC PDU sent by the LT.

ETP.3.1.1.1.2 Rejection of connection (INI/CON/VAL/REJ)

Ensure that the IUT transmits a CR PDU and reacts properly to the receipt of an M-RESETindication initiated by the LT.

ETP.3.1.1.2 Response to the receipt of invalid PDUs (INI/CON/INV/)

ETP.3.1.1.2.1 PDU with unknown PDU identifier (INI/CON/INV/UPI)

Ensure that the IUT transmits a CR PDU and reacts properly to the receipt of an invalid PDU with unknown PID sent by the LT.

ETP.3.1.1.2.2 PDU with unknown field (INI/CON/INV/UFI)

Ensure that the IUT transmits a CR PDU and reacts properly to the receipt of an invalid CC PDU with an unknown field sent by the LT.

ETP.3.1.1.2.3 PDU with missing field (INI/CON/INV/MFI)

Ensure that the IUT transmits a CR PDU and reacts properly to the receipt of an invalid CC PDU without the field I_ADDR, sent by the LT.

ETP.3.1.1.2.4 PDU with duplicated field (INI/CON/INV/DFI)

Ensure that the IUT transmits a CR PDU and reacts properly to the receipt of an invalid CC PDU with duplicated optional field WS, sent by the LT.

ETP.3.1.1.2.5 PDU with invalid order of fields (INI/CON/INV/RFI)

Ensure that the IUT transmits a CR PDU and reacts properly to the receipt of an invalid CC PDU with reordered fields R_ADDR and I_ADDR, sent by the LT.

ETP.3.1.1.2.6 PDU with field contents error (INI/CON/INV/FCE/)

Ensure that the IUT transmits a CR PDU and reacts properly to the receipt of an invalid CC PDU with an invalid WS field, sent by the LT.

Ensure that the IUT transmits a CR PDU and reacts properly to the receipt of an invalid CC PDU with a WS field longer than valid, sent by the LT.

Ensure that the IUT transmits a CR PDU and reacts properly to the receipt of an invalid CC PDU with a WS field shorter than valid, sent by the LT.

ETP.3.1.1.2.7 Response to inopportune PDUs (INI/CON/INV/INO)

Ensure that the IUT transmits a CR PDU and reacts properly to the receipt of an inopportune AC PDU sent by the LT.

ETP.3.1.2. Disconnection in the data transfer phase (INI/DIS/)

ETP.3.1.2.1 Response to the receipt of valid PDUs (INI/DIS/VAL/)

ETP.3.1.2.1.1 Initiator user disconnect (INI/DIS/VAL/IUD)

Ensure that the IUT is able to complete the user initiated disconnection in the data phase by sending a DR PDU with the appropriate reason and originator fields.

ETP.3.1.2.1.2 Responder disconnect (INI/DIS/VAL/RED)

Ensure that the IUT reacts properly to the receipt of a DR PDU in the data phase initiated by the LT.

ETP.3.1.2.2 Response to the receipt of invalid PDUs (INI/DIS/INV/)

ETP.3.1.2.2.1 PDU with unknown field (INI/DAT/INV/UFI)

Ensure that the IUT reacts properly to the receipt of an invalid DR PDU with an unknown field, sent by the LT in the data phase.

ETP.3.1.2.2.2 PDU with missing field (INI/DIS/INV/MFI)

Ensure that the IUT reacts properly to the receipt of an invalid DR PDU without the mandatory field I_ADDR, sent by the LT in the data phase.

ETP.3.1.2.2.3 PDU with duplicated field (INI/DIS/INV/DFI)

Ensure that the IUT reacts properly to the receipt of an invalid DR PDU with the duplicated field REA, sent by the LT.

ETP.3.1.2.2.4 PDU with reordered fields (INI/DIS/INV/RFI)

Ensure that the IUT reacts properly to the receipt of an invalid DR PDU with interchanged fields R_ADDR and I_ADDR, sent by the LT.

ETP.3.1.2.2.5 PDU with field contents error (INI/DIS/INV/FCE)

Ensure that the IUT reacts properly to the receipt of an invalid DR PDU with an invalid field I_ADDR, sent by the LT.

ETP.3.1.3 Data transfer (INI/DAT/)

ETP.3.1.3.1 Response to the receipt of valid PDUs (INI/DAT/VAL/)

ETP.3.1.3.1.1 Acknowledgement mechanism (INI/DAT/VAL/ACK/)

Ensure that the IUT transmits a negotiated number (NDAT) of DT PDUs in the data phase and reacts properly to the receipt of AC PDUs sent by the LT, where the EC was established by a CC without WS. This test purpose cannot be tested if WSR>1.

Ensure that the IUT transmits a negotiated number of DT PDUs (NDAT>WSR) in the data phase and reacts properly to the receipt of AC PDUs sent by the LT where the EC was established by a CC with WS.

ETP.3.1.3.2 Response to the receipt of invalid PDUs (INI/DAT/INV/)

ETP.3.1.3.2.1 PDU with unknown PDU identifier (INI/DAT/INV/UPI)

Ensure that the IUT reacts properly to the receipt of an invalid PDU with unknown PID, sent by the LT in the data phase.

ETP.3.1.3.2.2 PDU with unknown field (INI/DAT/INV/UFI)

Ensure that the IUT reacts properly to the receipt of an invalid AC PDU with an unknown field, sent by the LT in the data phase.

ETP.3.1.3.2.3 PDU with missing field (INI/DAT/INV/MFI)

Ensure that the IUT reacts properly to the receipt of an invalid AC PDU without the mandatory field I_ADDR, sent by the LT in the data phase.

ETP.3.1.3.2.4 PDU with duplicated field (INI/DAT/INV/DFI)

Ensure that the IUT transmits a negotiated number of DT PDUs in the data phase and reacts properly to the receipt of an invalid AC PDU with the duplicated field WS, sent by the LT.

ETP.3.1.3.2.5 PDU with reordered fields (INI/DAT/INV/RFI)

Ensure that the IUT reacts properly to the receipt of an invalid AC PDU with interchanged fields R_ADDR and I_ADDR, sent by the LT.

ETP.3.1.3.2.6 PDU with field contents error (INI/DAT/INV/FCE)

Ensure that the IUT transmits a negotiated number of DT PDUs in the data phase and reacts properly to the receipt of an invalid AC PDU with an invalid field I_ADDR, sent by the LT.

ETP.3.1.3.2.7 Behaviour after inopportune PDUs (INI/DAT/INV/INO)

Ensure that the IUT transmits a negotiated number of DT PDUs in the data phase and reacts properly to the receipt of an inopportune CC PDU, sent by the LT instead of a valid AC PDU.

ETP.3.2 Test purposes for responder IUTs.

— Exercise for the reader —

END OF DOCUMENT ETP

5.3 The EXPRO PICS

The proforma tables have already been completed with information for an example IUT. This information is written in italics.

DOCUMENT EPICS

This document provides the PICS proforma for the EXPRO protocol defined in EProt.

TABLE 5.8. EPICS.1 Initiator/responder capabilities

Item No.	Role	Reference	Status	Support
1	Initiator	EProt.6.1	o	*yes*
2	Responder	EProt.6.1	c1	*no*

init = EPICS.1/1 c1: IF init THEN x ELSE m

TABLE 5.9. EPICS.2 Supported PDUs

Item No.	**PDU**	**Sending of PDUs**			**Reception of PDUs**		
		Reference	**Status**	**Support**	**Reference**	**Status**	**Support**
1	CR	EProt.3.1	c1	*yes*	EProt.3.1	c2	
2	CC	EProt.3.1	c2		EProt.3.1	c1	*yes*
3	DR	EProt 3.2	m	*yes*	EProt 3.2	m	*yes*
4	DT	EProt.3.2	c1	*yes*	EProt.3.2	c2	
5	AC	EProt.3.2	c2		EProt.3.2	c1	*yes*

init = EPICS.1/1 resp = EPICS.1/2

c1: IF init THEN m ELSE n/a c2: IF resp THEN m ELSE n/a

TABLE 5.10. EPICS.3 Supported Parameters

Item No.	**Parameter**	**Reference**	**Status**	**Support**	**Values**		
					Status	**Allowed**	**Supported**
1	WS	EProt.4	c1	*yes*	c2	[1..255]	

resp = EPICS.1/2 withws = EPICS.3/1a

c1: IF resp THEN o ELSE m c2: IF resp AND withws THEN m ELSE n/a

END OF DOCUMENT EPICS

5.4 A PIXIT for EXATS

In the testing of an EXPRO responder IUT, the test oriented timer T_W is used to limit the time spent in a test case waiting for a response by the IUT. After any event supposed to trigger a reaction by the IUT, the event corresponding to this reaction is expected at the LT or UT within T_W time units. Depending on the situation, the waiting time may include IUT user response times and one or two medium transmission delays. The duration value of T_W is subject of agreement between clients and test laboratories. The time limit presumes a known bound for transmission delay in the medium, and a known bound for response times of the EXPRO entities EI and ER, as well as their users.

During conformance testing of EI or ER IUTs the only medium user entities are the IUT and the LT, and the only EXPRO user entities are the UT and the LT.

We have confined ourselves mainly to listing the parameter values needed for testing. We omitted sections dealing with identification and other administrative matters. The proforma tables have already been completed with information for an example test configuration. This information is written in italics.

DOCUMENT EPIX

This document provides the PIXIT proforma for the abstract test suite EXATS for the EXPRO protocol defined in EProt.

NEGOTIATED EXIT

The client and the test laboratory mutually agree to terminate the conformance assessment process when the state transitions from "idle_I" to "data_I" and vice versa of the initiator IUT versa cannot be performed.

TABLE 5.11. EPIX.1 Timer values

Timer Name	**Timer Range**	**Timer Value**	**Comment**
T_W	*[1..300] sec*	*10 sec*	This timer is used by the LT and UT to wait for IUT reactions.

TABLE 5.12. EPIX.2 Addresses

Parameter Name	**Parameter Type**	**PICS Clause**	**Parameter Range**	**Parameter Value**	**Comments**
I_ADDR	OCTETSTRING[1]		'01'O..'FF'O	*'01'O*	[1]
R_ADDR	OCTETSTRING[1]		'01'O..'FF'O	*'02'O*	[2]

[1] Value has to be provided by the client, if the initiator behaviour is to be tested, and by the test laboratory, if the responder behaviour is to be tested.

[2] Value has to be provided by the client, if the responder behaviour is to be tested and by the test laboratory, if the initiator behaviour is to be tested.

TABLE 5.13. EPIX.3 Parameter values

Parameter Name	**Parameter Type**	**Parameter Value**	**Comments**
NDAT	INTEGER	*100*	[1]

[1] Number of DTs to be sent or received. The value is negotiated between the test laboratory and the client.

END OF DOCUMENT EPIX

5.5 The EXPRO abstract test suite EXATS

Note 1: We took the liberty of omitting some unused table header comment lines.

Note 2: At the back of the test suite we are providing an identifier index. It can be quite helpful in reading the ATS.

DOCUMENT EXATS

5.5.1 *The test suite overview*

TABLE 5.14. Test suite structure

TEST SUITE STRUCTURE			
SUITE NAME : EXATS STANDARDS REF : EProt PICS REF : EPICS PIXIT REF : EPIX TEST METHOD(S) : Distributed COMMENTS : The test suite structure and test group references are derived from the TSS&TP document ETP.			
TEST GROUP REFERENCE	SELECTION REF	TEST GROUP OBJECTIVE	PAGE NR
INI/	IC	initiator capabilities	
INI/CON/		connection establishment	
INI/CON/VAL/		only valid PDUs are sent by the LT	
INI/CON/VAL/ACC/		LT accepts new connection	
INI/CON/VAL/REJ/		LT rejects new connection	
INI/CON/INV/		invalid PDUs are sent by the LT	
INI/CON/INV/UPI/		unknown protocol identifier	
INI/CON/INV/UFI/		unknown field	
INI/CON/INV/MFI/		missing field	
INI/CON/INV/DFI/		duplicated field	
INI/CON/INV/RFI/		reordered fields	
INI/CON/INV/FCE/		field contents error	
INI/CON/INV/INO/		inopportune PDU	
INI/DIS/		disconnect in data phase	
INI/DIS/VAL/		only valid PDUs are sent by the LT	
INI/DIS/VAL/IUD/		initiator user disconnect	
INI/DIS/VAL/RED/		responder disconnect	
INI/DIS/INV/		invalid PDUs are sent by the LT	
INI/DIS/INV/UFI/		unknown field	
INI/DIS/INV/MFI/		missing field	
INI/DIS/INV/DFI/		duplicated field	
INI/DIS/INV/RFI/		reordered fields	
INI/DIS/INV/FCE/		field contents error	
INI/DAT/		data transfer	
INI/DAT/VAL/		only valid PDUs are sent by the LT	
INI/DAT/VAL/ACK/		acknowledgement mechanism	
INI/DAT/INV/		invalid PDUs are sent by the LT	
INI/DAT/INV/UPI/		unknown protocol identifier	

Continued on next page

Continued from previous page

TEST GROUP REFERENCE	SELECTION REF	TEST GROUP OBJECTIVE	PAGE NR
INI/DAT/INV/UFI/		unknown field	
INI/DAT/INV/MFI/		missing field	
INI/DAT/INV/DFI/		duplicated field	
INI/DAT/INV/RFI/		reordered fields	
INI/DAT/INV/FCE/		field contents error	
INI/DAT/INV/INO/		inopportune PDU	
RES/	RC	responder capabilities, exercise	

TABLE 5.15 Test case index

TEST CASE INDEX				
TEST GROUP REFERENCE	TEST CASE ID	SEL REF	DESCRIPTION	PAGE NR
INI/CON/VAL/ACC/	TC_ICOV_ACC		accept EC	
INI/CON/VAL/REJ/	TC_ICOV_REJ		reject EC	
INI/CON/INV/UPI/	TC_ICOI_UPI		PDU with unknown PID	
INI/CON/INV/UFI/	TC_ICOI_UFI		CC with unknown field	
INI/CON/INV/MFI/	TC_ICOI_MFI		CC with missing field	
INI/CON/INV/DFI/	TC_ICOI_DFI		CC with duplicated fields	
INI/CON/INV/RFI/	TC_ICOI_RFI		CC with reordered fields	
INI/CON/INV/FCE/	TC_ICOI_FCE_1		CC with invalid WS field contents	
	TC_ICOI_FCE_2		CC with WS field length too long	
	TC_ICOI_FCE_3		CC with WS field length too short	
INI/CON/INV/INO/	TC_ICOI_INO		inopportune AC	
INI/DIS/VAL/IUD/	TC_IDIV_IUD		initiator user disconnect	
INI/DIS/VAL/RED/	TC_IDIV_RED		responder disconnect	
INI/DIS/INV/UFI/	TC_IDII_UFI		DR with unknown field	
INI/DIS/INV/MFI/	TC_IDII_MFI		DR with missing field	
INI/DIS/INV/DFI/	TC_IDII_DFI		DR with duplicated fields	
INI/DIS/INV/RFI/	TC_IDII_RFI		DR with reordered fields	
INI/DIS/INV/FCE/	TC_IDII_FCE		DR with invalid address	
INI/DAT/VAL/ACK/	TC_IDAV_ACK_1		CC without WS	
	TC_IDAV_ACK_2		CC with WS	
INI/DAT/INV/UPI/	TC_IDAI_UPI		PDU with unknown PID	
INI/DAT/INV/UFI/	TC_IDAI_UFI		AC with unknown field	
INI/DAT/INV/MFI/	TC_IDAI_MFI		AC with missing field	
INI/DAT/INV/DFI/	TC_IDAI_DFI		AC with duplicated fields	
INI/DAT/INV/RFI/	TC_IDAI_RFI		AC with reordered fields	
INI/DAT/INV/FCE/	TC_IDAI_FCE		AC with invalid address	
INI/DAT/INV/INO/	TC_IDAI_INO		inopportune CC	

TABLE 5.16. Test step index

TEST STEP INDEX			
TEST STEP GROUP REF.	TEST STEP ID	DESCRIPTION	PAGE NR
EXP/TSP/INI/	TS_IDLE_I_LT TS_IDLE_I_UT TS_DATA_I	LT to bring the initiator IUT to the state "idle_I" UT to bring the initiator IUT to the state "idle_I" to bring the initiator IUT to the state "data_I"	

TABLE 5.17. Default index

DEFAULT INDEX			
DEF GROUP REF	DEFAULT ID	DESCRIPTION	P NR
EXP/DEF/INI/	DAT_I	standard alternatives in the data phase	

5.5.2 The declarations part

In order to provide a clear naming convention for the TTCN types used in this test suite, the following prefixes in type names have been chosen:

SIT_ prefix used for simple types
STT_ prefix used for structured types
TSP_ prefix used for test suite parameters
TSV_ prefix used for test suite variables
TSC_ prefix used for test suite constants
TCV_ prefix used for test case variables

TABLE 5.18. Definition of simple types

SIMPLE TYPE DEFINITIONS		
TYPE NAME	TYPE DEFINITION	COMMENTS
SIT_PID	OCTETSTRING[1]	PDU identifier
SIT_ADD	OCTETSTRING[1]	EI/ER address
SIT_FIE	OCTETSTRING[1]	field identifier
SIT_LEN	OCTETSTRING[1]	field length
SIT_UD	OCTETSTRING[0..ESDUM]	user data
SIT_WS	INTEGER[1..255]	window size
SIT_REA	OCTETSTRING[1]	reason
SIT_ORI	OCTETSTRING[1]	originator
SIT_STATE	IA5String ("idle_I", "wait_CC", "data_I", "wait_AC", "idle_R", "wait_resp", "data_R", "wait_ready")	EI/ER states

TABLE 5.19. Definitions of structured types

STRUCTURED TYPE DEFINITION		
TYPE NAME :	STT_ADD	
COMMENTS :	The initiator and responder addresses, identifying the connection	
ELEMENT NAME	TYPE DEFINITION	COMMENTS
I_ADDR	SIT_ADD	initiator address
R_ADDR	SIT_ADD	responder address

STRUCTURED TYPE DEFINITION		
TYPE NAME :	STT_WS	
COMMENTS :	a structured type for the window size field	
ELEMENT NAME	TYPE DEFINITION	COMMENTS
WS_I	SIT_FIE	window size field identifier
WS_C	SIT_WS	window size field contents

STRUCTURED TYPE DEFINITION		
TYPE NAME :	STT_UD	
COMMENTS :	a structured type for the user data field	
ELEMENT NAME	TYPE DEFINITION	COMMENTS
UD_I	SIT_FIE	user data field identifier
UD_L	SIT_LEN	user data field length
UD_C	SIT_UD	user data field contents

TABLE 5.20. Definition of test suite operations

TEST SUITE OPERATION DEFINITION	
OPERATION NAME :	GEN_DAT(L:INTEGER)
RESULT TYPE :	OCTETSTRING[L]
DEFINITION	
GEN_DAT(L) returns an octetstring of length L. A negative L is treated as 0. Examples: GEN_DAT(-1)=''O, GEN_DAT(0)=''O, GEN_DAT(1)='00'O, '01'O, ..., or 'FF'O	
DETAILED COMMENTS : GEN_DAT is based on a fixed string OSTR of pseudo-random octets. OSTR is determined by the test laboratory and always used by it for this ATS. During the execution of EXATS, each call of GEN_DAT yields the first L octets of OSTR and then removes them from OSTR. Thus, although repetitions of identical calls of GEN_DAT usually deliver different values, the tester behavior is repeatable.	

TABLE 5.20 continued. Definition of test suite operations

TEST SUITE OPERATION DEFINITION		
OPERATION NAME	:	OCT_TO_INT(O:OCTETSTRING)
RESULT TYPE	:	INTEGER
DEFINITION		
OCT_TO_INT(O) converts OCTETSTRING to INTEGER. Example: OCT_TO_INT('01FF'O)=511		
DETAILED COMMENTS : It is analogous to the predefined conversion operation HEX_TO_INT		

TABLE 5.21. Declarations of test suite parameters

TEST SUITE PARAMETER DECLARATIONS			
PARAMETER NAME	PARAMETER TYPE	PICS/PIXIT REF	COMMENTS
TSP_IC	BOOLEAN	EPICS.1/1	initiator capabilities?
TSP_WSC	BOOLEAN	EPICS.3/1a	If IUT=ER: window size in CC?
TSP_WS	INTEGER	EPICS.3/1b	If IUT=ER and WS is supported: value of window size
TSP_W	INTEGER	EPIX.1/1	max. waiting time for IUT
TSP_I_ADDR	SIT_ADD	EPIX.2/1	initiator address
TSP_R_ADDR	SIT_ADD	EPIX.2/2	responder address
TSP_ND	INTEGER	EPIX.3/1	number of DTs to be sent

TABLE 5.22. Definitions of test case selection expressions

TEST CASE SELECTION EXPRESSION DEFINITIONS		
EXPRESSION NAME	SELECTION EXPRESSION	COMMENTS
IC	TSP_IC	initiator capabilities?
WSC	TSP_WSC	window size in CC?

TABLE 5.23. Declarations of test suite variables

TEST SUITE VARIABLE DECLARATIONS			
VARIABLE NAME	TYPE	VALUE	COMMENTS
TSV_ST_R	SIT_STATE	"idle_R"	LT state, used if IUT=EI
TSV_ST_I	SIT_STATE	"idle_I"	LT state, used if IUT=ER

TABLE 5.24. Declarations of test suite constants

TEST SUITE CONSTANT DECLARATIONS			
CONSTANT NAME	TYPE	VALUE	COMMENTS
TSC_MWS	OCTETSTRING[1]	'FF'O	maximal WSR
TSC_MESDUM	OCTETSTRING[1]	'FF'O	maximal ESDU length
TSC_CR_PID	SIT_PID	'00'O	CR PDU identifier
TSC_CC_PID	SIT_PID	'01'O	CC PDU identifier
TSC_DT_PID	SIT_PID	'02'O	DT PDU identifier
TSC_AC_PID	SIT_PID	'03'O	AC PDU identifier
TSC_DR_PID	SIT_PID	'04'O	DR PDU identifier
TSC_UD_FID	SIT_FIE	'00'O	user data field id.
TSC_WS_FID	SIT_FIE	'01'O	window size field id.
TSC_NORM	SIT_REA	'00'O	REA: normal
TSC_CONRF	SIT_REA	'01'O	REA: connection refusal
TSC_ERR	SIT_REA	'02'O	REA: procedure error
TSC_MRES	SIT_REA	'03'O	REA: medium reset
TSC_EIU	SIT_ORI	'00'O	ORI: initiator user
TSC_ERU	SIT_ORI	'01'O	ORI: responder user
TSC_EI	SIT_ORI	'02'O	ORI: initiator
TSC_ER	SIT_ORI	'03'O	ORI: responder
TSC_MED	SIT_ORI	'04'O	ORI: medium

TABLE 5.25. Declarations of test case variables

TEST CASE VARIABLE DECLARATIONS			
VARIABLE NAME	TYPE	VALUE	COMMENTS
TCV_UD	SIT_UD		user data contents field
TCV_VS	INTEGER[1..TSP_WS]		local send variable
TCV_SC	INTEGER		DT send counter
TCV_ST_I	SIT_STATE		assumed IUT state, used if IUT=EI

TABLE 5.26. Declarations of timers

TIMER DECLARATIONS			
TIMER NAME	DURATION	UNIT	COMMENTS
T_W	TSP_W	s	used to limit the waiting time for IUT reactions, gets its value from the PIXIT value EPIX.1/1b

TABLE 5.27. Declarations of PCOs

PCO DECLARATIONS			
PCO NAME	PCO TYPE	ROLE	COMMENTS
UI	ESAP	UT	EXPRO service access point between the upper tester and the initiator
UR	ESAP	UT	EXPRO service access point between the upper tester and the responder
LI	MSAP	LT	Medium service access point between the lower tester (in initiator role) and the medium
LR	MSAP	LT	Medium service access point between the lower tester (in responder role) and the medium

TABLE 5.28 Tabular ASP type definitions

ASP TYPE DEFINITION		
ASP NAME : CORQ(E-CONNECTrequest) PCO TYPE : UI		
PARAMETER NAME	PARAMETER TYPE	COMMENTS
<-	STT_ADD	initiator+responder addresses

ASP TYPE DEFINITION		
ASP NAME : COCO(E-CONNECTconfirm) PCO TYPE : UI		
PARAMETER NAME	PARAMETER TYPE	COMMENTS
<-	STT_ADD	initiator+responder addresses

ASP TYPE DEFINITION		
ASP NAME : DARY(E-DATAready) PCO TYPE : UI		
PARAMETER NAME	PARAMETER TYPE	COMMENTS
<-	STT_ADD	initiator+responder addresses

ASP TYPE DEFINITION		
ASP NAME : DARQ(E-DATArequest) PCO TYPE : UI		
PARAMETER NAME	PARAMETER TYPE	COMMENTS
<-	STT_ADD	initiator+responder addresses
ESDU	STT_UD	service data unit

TABLE 5.28 continued. Tabular ASP type definitions

ASP TYPE DEFINITION		
ASP NAME : DIRQ(E-DISCONNECTrequest) PCO TYPE : UI		
PARAMETER NAME	PARAMETER TYPE	COMMENTS
<-	STT_ADD	initiator+responder addresses
REA	OCTETSTRING[1]	reason
ORI	OCTETSTRING[1]	originator

ASP TYPE DEFINITION		
ASP NAME : DIIN(E-DISCONNECTindication) PCO TYPE : UI		
PARAMETER NAME	PARAMETER TYPE	COMMENTS
<-	STT_ADD	initiator+responder addresses
REA	OCTETSTRING[1]	reason
ORI	OCTETSTRING[1]	originator

ASP TYPE DEFINITION		
ASP NAME : MDRQ(M-DATArequest) PCO TYPE : LR		
PARAMETER NAME	PARAMETER TYPE	COMMENTS
MSDU	PDU	any EXPRO PDU

ASP TYPE DEFINITION		
ASP NAME : MDIN(M-DATAindication) PCO TYPE : LR		
PARAMETER NAME	PARAMETER TYPE	COMMENTS
MSDU	PDU	any EXPRO PDU

ASP TYPE DEFINITION		
ASP NAME : MRIN(M-RESETindication) PCO TYPE : LR		
PARAMETER NAME	PARAMETER TYPE	COMMENTS
REA	SIT_REA	reason
ORI	SIT_ORI	originator

TABLE 5.29. Tabular PDU type definitions

PDU TYPE DEFINITION		
PDU NAME : CR PCO TYPE : LR		
FIELD NAME	FIELD TYPE	COMMENTS
PID	SIT_PID	PDU identifier
<-	STT_ADD	initiator+responder addresses

PDU TYPE DEFINITION		
PDU NAME : CC PCO TYPE : LR		
FIELD NAME	FIELD TYPE	COMMENTS
PID	SIT_PID	PDU identifier
<-	STT_ADD	initiator+responder address
WS	STT_WS	window size field

PDU TYPE DEFINITION		
PDU NAME : DR PCO TYPE : LR		
FIELD NAME	FIELD TYPE	COMMENTS
PID	SIT_PID	PDU identifier
<-	STT_ADD	initiator+responder address
REA	SIT_REA	reason
ORI	SIT_ORI	originator

PDU TYPE DEFINITION		
PDU NAME : DT PCO TYPE : LR		
FIELD NAME	FIELD TYPE	COMMENTS
PID	SIT_PID	PDU identifier
<-	STT_ADD	initiator+responder address
UD	STT_UD	user data field

PDU TYPE DEFINITION		
PDU NAME : AC PCO TYPE : LR		
FIELD NAME	FIELD TYPE	COMMENTS
PID	SIT_PID	PDU identifier
<-	STT_ADD	initiator+responder address

TABLE 5.30. Alias definitions

ALIAS DEFINITIONS		
ALIAS NAME	EXPANSION	COMMENTS
CRin	MDIN	alias for the MDIN ASP used to carry a CR EPDU
CCout	MDRQ	alias for the MDRQ ASP used to carry a CC EPDU
DRout	MDRQ	alias for the MDRQ ASP used to carry a DR EPDU
DRin	MDRQ	alias for the MDIN ASP used to carry a DR EPDU
DTin	MDIN	alias for the MDIN ASP used to carry a DT EPDU
ACout	MDRQ	alias for the MDRQ ASP used to carry an AC EPDU
DETAILED COMMENTS: The alias names consist of a PDU name and a suffix "in" or "out", denoting PDUS received or sent by the LT.		

5.5.3 *The constraints part*

TABLE 5.31. Structured type constraint declarations

STRUCTURED TYPE CONSTRAINT DECLARATION		
CONSTRAINT NAME : ADDR STRUCTURED TYPE : STT_ADD DERIVATION PATH :		
ELEMENT NAME	ELEMENT VALUE	COMMENTS
I_ADDR	TSP_I_ADDR	test suite parameter
R_ADDR	TSP_R_ADDR	test suite parameter

STRUCTURED TYPE CONSTRAINT DECLARATION		
CONSTRAINT NAME : WIN STRUCTURED TYPE : STT_WS DERIVATION PATH : COMMENTS : a valid window size field with maximal window size		
ELEMENT NAME	ELEMENT VALUE	COMMENTS
WS_I	TSC_WS_FId	WS field identifier
WS_C	TSC_MWS	maximal WSR

STRUCTURED TYPE CONSTRAINT DECLARATION		
CONSTRAINT NAME : WIN_INV_WS_C STRUCTURED TYPE : STT_WS DERIVATION PATH : COMMENTS : a window size field with invalid contents subfield		
ELEMENT NAME	ELEMENT VALUE	COMMENTS
WS_I	TSC_WS_FId	WS field identifier
WS_C	'00'O	invalid contents

TABLE 5.31 continued. Structured type constraint declarations

STRUCTURED TYPE CONSTRAINT DECLARATION		
CONSTRAINT NAME : DATA		
STRUCTURED TYPE : STT_UD		
DERIVATION PATH :		
COMMENTS : a valid user data field with maximal length		
ELEMENT NAME	ELEMENT VALUE	COMMENTS
UD_I	TSC_UD_FId	UD field identifier
UD_L	TSC_MESDUM	maximal ESDU length
UD_C	TCV_UD	UD string

TABLE 5.32. Tabular ASP constraint declarations

ASP CONSTRAINT DECLARATION		
CONSTRAINT NAME : ECRQ		
ASP TYPE : CORQ		
DERIVATION PATH :		
COMMENTS : E-CONNECTrequest		
PARAMETER NAME	PARAMETER VALUE	COMMENTS
<-	ADDR	addresses

ASP CONSTRAINT DECLARATION		
CONSTRAINT NAME : ECCO		
ASP TYPE : COCO		
DERIVATION PATH :		
COMMENTS : E-CONNECTconfirm		
PARAMETER NAME	PARAMETER VALUE	COMMENTS
<-	ADDR	addresses

ASP CONSTRAINT DECLARATION		
CONSTRAINT NAME : EDAR(X:SIT_UD)		
ASP TYPE : DARQ		
DERIVATION PATH :		
COMMENTS : E-DATArequest, parameterized		
PARAMETER NAME	PARAMETER VALUE	COMMENTS
<-	ADDR	addresses
ESDU	X	formal parameter user data

TABLE 5.32 continued (1). Tabular ASP constraint declarations

ASP CONSTRAINT DECLARATION		
CONSTRAINT NAME : EDRY		
ASP TYPE : DARY		
DERIVATION PATH :		
COMMENTS : E-DATAready		
PARAMETER NAME	PARAMETER VALUE	COMMENTS
<-	ADDR	addresses

ASP CONSTRAINT DECLARATION		
CONSTRAINT NAME : EDRQ(R:SIT_REA;O:SIT_ORI)		
ASP TYPE : DIRQ		
DERIVATION PATH :		
COMMENTS : E-DISCONNECTrequest, parameterized		
PARAMETER NAME	PARAMETER VALUE	COMMENTS
<-	ADDR	addresses
REA	R	formal parameter reason
ORI	O	formal parameter originator

ASP CONSTRAINT DECLARATION		
CONSTRAINT NAME : EDIN(R:SIT_REA;O:SIT_ORI)		
ASP TYPE : DIIN		
DERIVATION PATH :		
COMMENTS : E-DISCONNECTindication, parameterized		
PARAMETER NAME	PARAMETER VALUE	COMMENTS
<-	ADDR	addresses
REA	R	formal parameter reason
ORI	O	formal parameter originator

ASP CONSTRAINT DECLARATION		
CONSTRAINT NAME : MDR(X:PDU)		
ASP TYPE : MDRQ		
DERIVATION PATH :		
COMMENTS : M-DATArequest, parameterized		
PARAMETER NAME	PARAMETER VALUE	COMMENTS
MSDU	X	formal parameter MSDU

TABLE 5.32 continued (2). Tabular ASP constraint declarations

ASP CONSTRAINT DECLARATION		
CONSTRAINT NAME : MDI(X:PDU) ASP TYPE : MDIN DERIVATION PATH : COMMENTS : M-DATAindication, parameterized		
PARAMETER NAME	PARAMETER VALUE	COMMENTS
MSDU	X	formal parameter MSDU

ASP CONSTRAINT DECLARATION		
CONSTRAINT NAME : MRI(R:SIT_REA;O:SIT_ORI) ASP TYPE : MRIN DERIVATION PATH : COMMENTS : M-RESETindication, parameterized		
PARAMETER NAME	PARAMETER VALUE	COMMENTS
REA	R	formal parameter reason
ORI	O	formal parameter originator

TABLE 5.33. Tabular PDU constraint declarations

PDU CONSTRAINT DECLARATION		
CONSTRAINT NAME : CR1 PDU TYPE : CR DERIVATION PATH : COMMENTS : valid CR PDU from IUT to LT		
FIELD NAME	FIELD VALUE	COMMENTS
PID	TSC_CR_PID	CR PDU identifier
<-	ADDR	addresses

PDU CONSTRAINT DECLARATION		
CONSTRAINT NAME : CC1 PDU TYPE : CC DERIVATION PATH : COMMENTS : valid CC PDU from LT to IUT		
FIELD NAME	FIELD VALUE	COMMENTS
PID	TSC_CC_PID	PDU identifier
<-	ADDR	addresses
WS	WIN	structured type constraint

TABLE 5.33 continued. Tabular PDU constraint declarations

PDU CONSTRAINT DECLARATION		
CONSTRAINT NAME :	CC2	
PDU TYPE :	CC	
DERIVATION PATH :		
COMMENTS :	invalid CC PDU from LT to IUT	
FIELD NAME	FIELD VALUE	COMMENTS
PID	TSC_CC_PID	PDU identifier
<-	ADDR	addresses
WS	WIN_INV_WS_C	invalid window size contents

PDU CONSTRAINT DECLARATION		
CONSTRAINT NAME :	DR1(RE:SIT_REA, OR:SIT_ORI)	
PDU TYPE :	DR	
DERIVATION PATH :		
COMMENTS :	valid DR PDU from LT to IUT or from IUT to LT	
FIELD NAME	FIELD VALUE	COMMENTS
PID	TSC_DR_PID	PDU identifier
<-	ADDR	addresses
REA	RE	reason, formal parameter
ORI	OR	originator, formal parameter

PDU CONSTRAINT DECLARATION		
CONSTRAINT NAME :	DT1	
PDU TYPE :	DT	
DERIVATION PATH :		
COMMENTS :	valid DT PDU from IUT to LT	
FIELD NAME	FIELD VALUE	COMMENTS
PID	TSC_DT_PID	PDU identifier
<-	ADDR	addresses
UD	DATA	structured type constraint

PDU CONSTRAINT DECLARATION		
CONSTRAINT NAME :	AC1	
PDU TYPE :	AC	
DERIVATION PATH :		
COMMENTS :	valid AC PDU from IUT to LT	
FIELD NAME	FIELD VALUE	COMMENTS
PID	TSC_AC_PID	PDU identifier
<-	ADDR	addresses

5.5.4 The dynamic part

TABLE 5.34. Definition of the test case TC_ICOV_ACC

TEST CASE DYNAMIC BEHAVIOUR		
TEST CASE NAME	:	TC_ICOV_ACC
GROUP	:	EXP/INI/CON/VAL/ACC/
PURPOSE	:	Ensure that the IUT transmits a CR PDU and reacts properly to the receipt of a CC PDU
DEFAULT	:	
COMMENTS	:	The test step TS_DATA_I is used to bring the initiator IUT from the state "idle_I" to the state "data_I"

LABEL	BEHAVIOUR DESCRIPTION	CONSTRAINTS REF	VERDICT	COMMENTS
	+TS_DATA_I			
	[R=fail]		F	
	[R=inconc]		I	
	[R=pass]		(P)	
	+TS_IDLE_I_UT			

TABLE 5.35. Definition of the test case TC_ICOI_FCE_1

TEST CASE DYNAMIC BEHAVIOUR		
TEST CASE NAME	:	TC_ICOI_FCE_1
GROUP	:	EXP/INI/CON/INV/FCE/
PURPOSE	:	Ensure that the IUT transmits a CR PDU and reacts properly to the receipt of an CC PDU with an invalid WS parameter (invalid contents field)
DEFAULT	:	
COMMENTS	:	The test step TS_IDLE_I_LT is used by the LT to bring the IUT back to the state "idle_I"

L	BEHAVIOUR DESCRIPTION	CONSTRAINTS REF	V
	UI!CORQ START T_W	ECRQ	
	LR?CRin CANCEL T_W	MDI(CR1)	
	LR!CCout START T_W	MDR(CC2)	
	LR?DRin	MDI(DR1(TSC_ERR,TSC_EI))	(P)
	UI?DIIN CANCEL T_W	EDIN(TSC_ERR,TSC_EI)	P
	UI?OTHERWISE CANCEL T_W		F
	?TIMEOUT T_W		F
	LR?MRIN	MRI(?,TSC_MED)	(I)
	UI?DIIN CANCEL T_W	EDIN(?,TSC_MED)	I
	UI?OTHERWISE CANCEL T_W		F
	?TIMEOUT T_W		F

[0] [4]

Continued on next page

Continued from previous page

[1] [3]

L	BEHAVIOUR DESCRIPTION	CONSTRAINTS REF	V
	LR?OTHERWISE CANCEL T_W		(F)
	+TS_IDLE_I_LT		
	?TIMEOUT T_W		(F)
	+TS_IDLE_I_LT		
	LR?MRIN CANCEL T_W	MRI(?,TSC_MED)	(I)
	UI?DIIN CANCEL T_W	EDIN(?,TSC_MED)	I
	UI?OTHERWISE CANCEL T_W		F
	?TIMEOUT T_W		F
	LR?OTHERWISE CANCEL T_W		F
	?TIMEOUT T_W		F

TABLE 5.36. Definition of the test case TC_IDAV_ACK_2

TEST CASE DYNAMIC BEHAVIOUR		
TEST CASE NAME	:	TC_IDAV_ACK_2
GROUP	:	EXP/INI/DAT/VAL/ACK/
PURPOSE	:	Ensure that the IUT transmits a previously negotiated number of DT PDUs in the data phase and reacts properly to the receipt of an AC PDU sent by the LT. The EC was established by CC with the optional parameter WS.
DEFAULT	:	DAT_I
COMMENTS	:	The test step TS_DATA_I is used to bring the initiator IUT from the state "idle_I" to the state "data_I".

L	BEHAVIOUR DESCRIPTION	CONSTRAINTS REF	V
	+TS_DATA_I		
	[R=fail]		F
	[R=inconc]		I
	[R=pass] (TCV_SC:=0,		
	#		
	TCV_VS:=OCT_TO_INT(TSC_MWS))		
	REPEAT SEND(TCV_SC) UNTIL [TCV_SC=TSP_ND]		
	+TS_IDLE_I_LT		
	SEND(S:INTEGER)		
	(TCV_UD:=GENDAT(TSC_MESDUM)		
	UI!DARQ START T_W (TCV_VS:=TCV_VS-1)	EDAR(TCV_UD)	
	[TCV_VS>0]		
	UI?DARY(S:=S+1)	EDRY	

[0] [3]

Continued on next page

Continued from previous page

[2] [4]

L	BEHAVIOUR DESCRIPTION	CONSTRAINTS REF	V
	LR?DTin CANCEL T_W	MDI(DT1)	(P)
	[TCV_VS=0]		
	LR?DTin CANCEL T_W	MDI(DT1)	(P)
	LR!ACout START T_W	MDR(AC1)	
	UI?DARY CANCEL T_W	EDRY	
	# (S:=S+1,TCV_VS:=TSC_MWS)		

TEST STEP LIBRARY

TABLE 5.37. Definition of the test step TS_DATA_I

TEST STEP DYNAMIC BEHAVIOUR		
TEST STEP NAME	:	TS_DATA_I
GROUP	:	EXP/STEP/RES/
OBJECTIVE	:	To bring the initiator IUT from the state "idle_I" to "data_I".
COMMENTS	:	The test step TS_IDLE_I_UT may be used by the UT to bring the initiator IUT to the state "idle_I". The test step TS_IDLE_I_LT may be used by the LT to bring the initiator IUT to the state "idle_I".

L	BEHAVIOUR DESCRIPTION	CONSTRAINTS REF	V
	UI!CORQ START T_W	ECRQ	
	LR?CRin CANCEL T_W	MDI(CR1)	
	LR!CCout START T_W (TSV_ST_I :="data_I")	MDR(CC1)	
	UI?COCO CANCEL T_W	ECCO	(P)
	UI?DIIN	EDIN(?,TSC_MED)	(I)
	LR?MRIN CANCEL T_W (TSV_ST_I:="idle_I")	MRI(?,TSC_MED)	I
	LR?OTHERWISE CANCEL T_W		(F)
	?TIMEOUT T_W		(F)
	UI?OTHERWISE CANCEL T_W		F
	LR?MRIN	MRI(?,TSC_MED)	(I)
	UI?DIIN CANCEL T_W	EDIN(?,TSC_MED)	
	UI?OTHERWISE CANCEL T_W		F
	?TIMEOUT T_W		
	+TS_IDLE_I_LT		
	LR?OTHERWISE CANCEL T_W		
	+TS_IDLE_I_LT		
	?TIMEOUT T_W		(F)

[0] [3]

Continued on next page

Continued from previous page

[1] [4]

L	BEHAVIOUR DESCRIPTION	CONSTRAINTS REF	V
	+TS_IDLE_I_LT		
	LR?MRIN (TSV_ST_I:="idle_I")	MRI(?TSC_MED)	(I)
	UI?DIIN CANCEL T_W	EDIN(?,TSC_MED)	(I)
	UI?OTHERWISE		(F)
	?TIMEOUT T_W		(F)
	LR?OTHERWISE CANCEL T_W		(F)
	?TIMEOUT T_W		(F)

TABLE 5.38. Definition of the test step TS_IDLE_I_UT

TEST STEP DYNAMIC BEHAVIOUR			
TEST STEP NAME : TS_IDLE_I_UT			
GROUP : EXP/STEP/INI/			
OBJECTIVE : Disconnect initiated by the UT to bring the initiator IUT from the state "data_I" to the state "idle_I".			
L	BEHAVIOUR DESCRIPTION	CONSTRAINTS REF	VERDICT
	[TSV_ST_I = "data_I"]		
	UI!DIRQ START T_W	EDRQ(TSC_NORM ,TSC_EIU)	
	# (TSV_ST_I = "idle_I")		
	LR?DRin CANCEL T_W	MDI(DR1(TSC_NO RM,TSC_EIU))	
	#		
	LR?OTHERWISE CANCEL T_W		
	?TIMEOUT T_W		

TABLE 5.39. Definition of the test step TS_IDLE_I_LT

TEST STEP DYNAMIC BEHAVIOUR			
TEST STEP NAME : TS_IDLE_I_LT			
GROUP : EXP/STEP/INI/			
OBJECTIVE : Disconnect initiated by the LT to bring the initiator IUT from the state "data_I" to the state "idle_I".			
L	BEHAVIOUR DESCRIPTION	CONSTRAINTS REF	V
	[TSV_ST_I = "data_I"]		
	LR!DRout START T_W	MDR(DR1(TSC_ERR,TSC_ER)	
	# (TSV_ST_I = "idle_I")		
	UI?DIIN CANCEL T_W	EDIN(TSC_ERR,TSC_ER)	
	UI?OTHERWISE CANCEL T_W		
	?TIMEOUT T_W		

DEFAULTS LIBRARY

TABLE 5.40. Definition of the default DAT_I

DEFAULT DYNAMIC BEHAVIOUR			
DEFAULT NAME : DAT_I			
GROUP : EXP/DEF/INI/DAT_I			
OBJECTIVE : handling of medium initiated resets, T_W timeouts and invalid			
COMMENTS : PDUs in the data phase. The IUT is initiator.			
LABEL	BEHAVIOUR DESCRIPTION	CONSTRAINTS REF	VERDICT
	UI?DIIN	EDIN(?,TSC_MRES)	(I)
	LR?MRIN CANCEL T_W	MRI(?,TSC_MRES)	I
	LR?OTHERWISE CANCEL T_W		(F)
	+TS_L_IDLE_I_LT		
	?TIMEOUT T_W		(F)
	+TS_L_IDLE_I_LT		
	UI?OTHERWISE CANCEL T_W		(F)
	+TS_L_IDLE_I_LT		
	LR?MRIN	MRI(?,TSC_MRES)	(I)
	UI?DIIN CANCEL T_W	EDIN(?,TSC_MRES)	I
	UI?OTHERWISE CANCEL T_W		(F)
	+TS_L_IDLE_I_LT		
	?TIMEOUT T_W		(F)
	LR?OTHERWISE		(F)
	+TS_L_IDLE_I_LT		
	?TIMEOUT T_W		(F)
	+TS_L_IDLE_I_LT		
	TS_L_IDLE_I_LT		
	[TSV_ST_I = "data_I"]		
	LR!DRout START T_W	MDR(DR1(TSC_ERR,	
	# (TSV_ST_I = "idle_I")	TSC_ER)	
	UI?DIIN CANCEL T_W	EDIN(TSC_ERR,TSC_ER)	R
	UI?OTHERWISE CANCEL T_W		R
	?TIMEOUT T_W		R

5.5.5 Identifier index

TABLE 5.41. Identifier index

IDENTIFIER	OBJECT	TAB.	IDENTIFIER	OBJECT	TAB.
AC	PDU type	5.29	SIT_LEN	simple type	5.18
AC1	PDU type constr.	5.33	SIT_ORI	simple type	5.18
ACout	alias	5.30	SIT_PID	simple type	5.18
ADDR	str. type constr.	5.31	SIT_REA	simple type	5.18
CC	PDU type	5.29	SIT_UD	simple type	5.18
CC1	PDU type constr.	5.33	SIT_WS	simple type	5.18
CC2	PDU type constr.	5.33	STT_ADD	struct. type	5.19
CCout	alias	5.30	STT_UD	struct. type	5.19
COCO	ASP type	5.28	STT_WS	struct. type	5.19
CORQ	ASP type	5.28	TCV_SC	test case variable	5.25
CR	PDU type	5.29	TCV_ST_I	test case variable	5.25
CR1	PDU type constr.	5.33	TCV_UD	test case variable	5.25
CRin	alias	5.30	TCV_VS	test case variable	5.25
DARQ	ASP type	5.28	TC_ICOI_FCE_1	test case	5.15
DARY	ASP type	5.28	TC_ICOV_ACC	test case	5.15
DATA	str. type constr	5.31	TC_IDAV_ACK_2	test case	5.15
DAT_I	default	5.17	TSC_AC_PID	test suite const.	5.24
DIIN	ASP type	5.28	TSC_CC_PID	test suite const.	5.24
DIRQ	ASP type	5.28	TSC_CR_PID	test suite const.	5.24
DR	PDU type ident.	5.29	TSC_DR_PID	test suite const.	5.24
DR1	PDU type constr.	5.33	TSC_DT_PID	test suite const.	5.24
DRin	alias	5.30	TSC_EI	test suite const.	5.24
DRout	alias	5.30	TSC_EIU	test suite const.	5.24
DT	PDU type	5.29	TSC_ER	test suite const.	5.24
DT1	PDU type constr.	5.33	TSC_ERR	test suite const.	5.24
DTin	alias	5.30	TSC_MED	test suite const.	5.24
ECCO	ASP type constr.	5.32	TSC_MESDUM	test suite const.	5.24
ECRQ	ASP type constr.	5.32	TSC_MRES	test suite const.	5.24
EDAR	ASP type constr.	5.32	TSC_MWS	test suite const.	5.24
EDIN	ASP type constr.	5.32	TSC_NORM	test suite const	5.24
EDRQ	ASP type constr.	5.32	TSC_UD_FID	test suite const..	5.24
EDRY	ASP type constr.	5.32	TSC_WS_FID	test suite const.	5.24
GEN_DAT	test suite oper.	5.20	TSP_I_ADDR	test suite param.	5.21
LR	PCO	5.27	TSP_ND	test suite param.	5.21
MDI	ASP type constr.	5.32	TSP_R_ADDR	test suite param.	5.21
MDR	ASP type constr.	5.32	TSV_ST_I	test suite variable	5.23
MRI	ASP type constr.	5.32	TS_DATA_I	test step	5.37
MRIN	ASP type	5.28	TS_IDLE_I_LT	test step	5.39
OCT_TO_INT	test suite oper.	5.20	TS_IDLE_I_UT	test step	5.38
SIT_ADD	simple type	5.18	TS_L_IDLE_I_LT	local test step	5.40
SIT_FIE	simple type	5.18	T_W	timer	5.26

TABLE 5.41. Identifier index, continued

IDENTIFIER	OBJECT	TAB.	IDENTIFIER	OBJECT	TAB.
UI	PCO	5.27	WIN_INV_WS_C	str. type constr.	5.31
WIN	str. type constr.	5.31			

END OF DOCUMENT EXATS

ANNEX A. THE BNF RULES FOR TTCN

This annex provides the BNF rules for TTCN in alphabetic order. The following notation is used in the TTCN BNF rules:

::=	production symbol
|	alternative
[...]	optional
{...}	zero or more instances
{...}+	one or more instances
(...)	grouping
xyz	non terminal symbol
xyz	terminal string (used for alphanumeric strings)
"xyz"	terminal string (used for special characters)
/* ...*/	static semantic rules

The BNF rule set presented here is an interim version that takes into account the amendments 1 and 2. The forthcoming second version of TTCN will probably differ very little, if at all, from it. The meaning of BNF is explained in Subsection 3.2.1.

Activate ::= ***ACTIVATE***"("[*DefaultRefList*]")"

/* The ACTIVATE construct shall not be used in Default behaviour tables. */

ActualCrefPar ::= *Value* | *DataObjectIdentifier* | *ConsRef*

ActualCrefParList ::= "(" *ActualCrefPar* {*Comma ActualCrefPar* } ")"

/* See static semantics on production for ActualParList */

ActualPar ::= *Value* | *PCO_Identifier* | *CP_Identifier*

ActualParList ::= "(" *ActualPar* {*Comma ActualPar* } ")"

/* The number of the actual parameters shall be the same as the number of the formal parameters. Each actual parameter shall resolve to a specific value compatible with the type of its corresponding formal parameter. If a parameter is a parameterized constraint then the constraint shall be passed together with its actual parameter list. The actual parameters shall be bound. If the type of the formal parameter is PDU, then the actual parameter's type shall be declared as PDU or as a specific PDU type */

AddOp ::= "+" | "-" | ***OR***

/* Operands of the "+", "-" operators shall be of type INTEGER (i.e., TTCN or ASN.1 predefined) or derivations of INTEGER (i.e., subrange). Operands of the OR operator shall be of type BOOLEAN (TTCN or ASN.1 predefined) or derivatives of BOOLEAN */

AliasDef ::= ***$AliasDef*** *AliasId ExpandedId* [*Comment*] ***$End_AliasDef***

AliasDefs ::= ***$Begin_AliasDefs*** *{AliasDef}+ [Comment]* ***$End_AliasDefs***

AliasId ::= ***$AliasId*** *AliasIdentifier*

AliasIdentifier ::= *Identifier*

/* An AliasIdentifier shall be used only in a statement line of a behaviour description. An AliasIdentifier shall be used only where an ASP_Identifier or PDU_Identifier is valid */

Alpha ::= *UpperAlpha | LowerAlpha*

AlphaNum ::= *Alpha | Num*

AnyOne ::= "?"

/* AnyOne shall be used only within values of string types, SEQUENCE OF and SET OF */

AnyOrNone ::= "*"

/* AnyOrNone shall be used only within values of string types, SEQUENCE OF and SET OF */

AnyOrOmit ::= "*"

AnyValue ::= "?"

ArrayRef ::= *Dot* "[" *ComponentNumber* "]"

ASN1_ASP_Constraint ::= ***$Begin_ASN1_ASP_Constraint*** *ConsId ASP_Id DerivPath [Comment] ASN1_ConsValue [Comment]* ***$End_ASN1_ASP_Constraint***

/* The FullIdentifier that is part of ASP_Id shall not be used . If an ASP is substructured, then the constraints for ASPs of that type shall have a compatible ASN.1 structure (i.e., possibly with some groupings).A modified constraint shall have the same parameter list as its base constraint. In particular, there shall be no parameters omitted from or added to this list */

ASN1_ASP_Constraints ::= ***$ASN1_ASP_Constraints*** *{ASN1_ASP_Constraint}+* ***$End_ASN1_ASP_Constraints***

ASN1_ASP_TypeDef ::= ***$Begin_ASN1_ASP_TypeDef*** *ASP_Id PCO_Type [Comment] ASN1_TypeDefinition [Comment]* ***$End_ASN1_ASP_TypeDef***

ASN1_ASP_TypeDefByRef ::= ***$ASN1_ASP_TypeDefByRef*** *ASP_Id PCO_Type ASN1_TypeReference ASN1_ModuleId [Comment]* ***$End_ASN1_ASP_TypeDefByRef***

/* ASP_Id shall not be specified with a FullIdentifier */

ASN1_ASP_TypeDefs ::= ***$ASN1_ASP_TypeDefs*** *{ASN1_ASP_TypeDef}* ***$End_ASN1_ASP_TypeDefs***

ASN1_ASP_TypeDefsByRef ::= ***$Begin_ASN1_ASP_TypeDefsByRef*** *{ASN1_ASP_TypeDefByRef}+ [Comment]* ***$End_ASN1_ASP_TypeDefsByRef***

ASN1_CM_Constraint ::= ***$Begin_ASN1_CM_Constraint*** *ConsId&Parlist CM_Id [DerivationPath] [Comment] ASN1_ConsValue [Comment]* ***$End_ASN1_CM_Constraint***

/* STATIC SEMANTICS - If a CM is substructured, then the constraints for CMs of that type shall have a compatible ASN.1 structure (i.e., possibly with some groupings). A modified constraint shall have the same parameter list as its base constraint. In particular, there shall be no parameters omitted from or added to this list */

ASN1_CM_Constraints ::= ***$ASN1_CM_Constraints*** *{ASN1_CM_Constraint}+* ***$End_ASN1_CM_Constraints***

ASN1_CM_TypeDef ::= ***$Begin_ASN1_CM_TypeDef*** *CM_Id* [*Comment*] *ASN1_TypeDefinition* [*Comment*] ***$End_ASN1_CM_TypeDef***

ASN1_CM_TypeDefs ::= ***$ASN1_CM_TypeDefs*** {*ASN1_CM_TypeDef*}+ ***$End_ASN1_CM_TypeDefs***

ASN1_ConsValue ::= ***$ASN1_ConsValue*** *ConstraintValue&AttributesOrReplace* ***$End_ASN1_ConsValue***

ASN1_Identifier ::= *Identifier*

/* ASN1_Identifier identifies a field within ASN.1 SEQUENCE, SET or CHOICE type. An ASN1_Identifier associated with a NamedValue shall not be used unless the value is within a SEQUENCE, SET or CHOICE type. An ASN1_Identifier shall be provided to identify the variant in a CHOICE type. An ASN1_Identifier shall be provided whenever the value definition becomes ambiguous because of omitted OPTIONAL values in a SEQUENCE type */

ASN1_LocalType ::= <u>*Typeassignment*</u>

/* Where Typeassignment is a non-terminal defined in ISO/IEC 8824. ASN.1 type definitions used within TTCN shall not use external type references as defined in ISO/IEC 8824 */

ASN1_ModuleId ::= ***$ASN1_ModuleId*** *ModuleIdentifier*

ASN1_PDU_Constraint ::= ***$Begin_ASN1_PDU_Constraint*** *ConsId PDU_Id DerivPath* [*Comment*] *ASN1_ConsValue* [*Comment*] ***$End_ASN1_PDU_Constraint***

/* The FullIdentifier that is part of PDU_Id shall not be used. If a PDU is substructured, then the constraints for PDUs of that type shall have a compatible ASN.1 structure (i.e., possibly with some groupings). A modified constraint shall have the same parameter list as its base constraint. In particular, there shall be no parameters omitted from or added to this list */

ASN1_PDU_Constraints ::= ***$ASN1_PDU_Constraints*** {*ASN1_PDU_Constraint*}+ **$End_ASN1_PDU_Constraints**

ASN1_PDU_TypeDef ::= ***$Begin_ASN1_PDU_TypeDef*** *PDU_Id PCO_Type* [*Comment*] *ASN1_TypeDefinition* [*Comment*] ***$End_ASN1_PDU_TypeDef***

/* If a PDU is sent or received only embedded in ASPs within the whole test suite, then PCO_TypeIdentifier (in PCO_Type) is optional */

ASN1_PDU_TypeDefByRef ::= ***$ASN1_PDU_TypeDefByRef*** *PDU_Id PCO_Type ASN1_TypeReference ASN1_ModuleId* [*Comment*] ***$End_ASN1_PDU_TypeDefByRef***

/* If a PDU is sent or received only embedded in ASPs within the whole test suite, then PCO_TypeIdentifier (in PCO_Type) is optional.PDU_Id shall not be specified with a FullIdentifier */

ASN1_PDU_TypeDefs ::= ***$ASN1_PDU_TypeDefs*** {*ASN1_PDU_TypeDef*} ***$End_ASN1_PDU_TypeDefs***

ASN1_PDU_TypeDefsByRef ::= ***$Begin_ASN1_PDU_TypeDefsByRef*** {*ASN1_PDU_TypeDefByRef*}+ [*Comment*] ***$End_ASN1_PDU_TypeDefsByRef***

ASN1_Type ::= <u>*Type*</u>

/* Where Type is a non-terminal defined in ISO/IEC 8824. Each terminal type reference used within the Type production shall be one of the following: ASN1_LocalType typereference, TS_TypeIdentifier or PDU_-Identifier. ASN.1 type definitions used within TTCN shall not use external type references as defined in ISO/IEC 8824 */

ASN1_Type&LocalTypes ::= *ASN1_Type* {*ASN1_LocalType*}

/* Types referred to from the ASN1_Type definition shall be defined in other ASN.1 type definition tables, be defined by reference in the ASN.1 type reference table or be defined locally (i.e., ASN1_LocalTypes) in the same table, following the first type definition. ASN1_LocalTypes shall not be used in other parts of the test suite */

ASN1_TypeConstraint ::= ***$Begin_ASN1_TypeConstraint*** *ConsId ASN1_TypeId DerivPath* [*Comment*] *ASN1_ConsValue* [*Comment*] ***$End_ASN1_TypeConstraint***

/* The FullIdentifier that is part of ASN1_TypeId shall not be used. A modified constraint shall have the same parameter list as its base constraint. In particular, there shall be no parameters omitted from or added to this list */

ASN1_TypeConstraints ::= ***$ASN1_TypeConstraints*** {*ASN1_TypeConstraint*}+ ***$End_ASN1_TypeConstraints***

ASN1_TypeDef ::= ***$Begin_ASN1_TypeDef*** *ASN1_TypeId* [*Comment*] *ASN1_TypeDefinition* [*Comment*] ***$End_ASN1_TypeDef***

ASN1_TypeDefinition ::= ***$ASN1_TypeDefinition*** *ASN1_Type&LocalTypes* ***$End_ASN1_TypeDefinition***

ASN1_TypeDefs ::= ***$ASN1_TypeDefs*** {*ASN1_TypeDef*}+ ***$End_ASN1_TypeDefs***

ASN1_TypeId ::= ***$ASN1_TypeId*** *ASN1_TypeId&FullId*

ASN1_TypeId&FullId ::= *ASN1_TypeIdentifier* [*FullIdentifier*]

ASN1_TypeIdentifier ::= *Identifier*

ASN1_TypeRef ::= ***$ASN1_TypeRef*** *ASN1_TypeId ASN1_TypeReference ASN1_ModuleId* [*Comment*] ***$End_ASN1_TypeRef***
/* ASN1_TypeId shall not be specified with a FullIdentifier */

ASN1_TypeReference ::= ***$ASN1_TypeReference*** *TypeReference*

ASN1_TypeRefs ::= ***$Begin_ASN1_TypeRefs*** {*ASN1_TypeRef*}+ [*Comment*] ***$End_ASN1_TypeRefs***

ASP_Constraints ::= ***$ASP_Constraints*** [*TTCN_ASP_Constraints*] [*ASN1_ASP_Constraints*] ***$End_ASP_Constraints***

ASP_Id ::= ***$ASP_Id*** *ASP_Id&FullId*

ASP_Id&FullId ::= *ASP_Identifier* [*FullIdentifier*]

ASP_Identifier ::= *Identifier*

ASP_ParDcl ::= ***$ASP_ParDcl*** *ASP_ParId ASP_ParType* [*Comment*] ***$End_ASP_ParDcl***

ASP_ParDcls ::= ***$ASP_ParDcls*** {*ASP_ParDcl*} ***$End_ASP_ParDcls***

ASP_ParId ::= ***$ASP_ParId*** *ASP_ParIdOrMacro*

ASP_ParId&FullId ::= *ASP_ParIdentifier* [*FullIdentifier*]

ASP_ParIdentifier ::= *Identifier*

ASP_ParIdOrMacro ::= *ASP_ParId&FullId* | *MacroSymbol*

/* The MacroSymbol shall be used only in combination with a reference to a Structured Type */

ASP_ParType ::= ***$ASP_ParType*** *Type&Attributes*

/* Type shall be a PredefinedType or TS_TypeIdentifier, PDU_Identifier, or PDU */

ASP_ParValue ::= ***$ASP_ParValue*** *ASP_ParId ConsValue* [*Comment*] ***$End_ASP_ParValue***

/* The FullIdentifier that is part of ASP_ParId shall not be used. If an ASP definition refers to a Structured Type as a substructure of a parameter (i.e., with a parameter name) then the corresponding constraint shall have the same parameter name in the corresponding position in the parameter name name column of the constraint and the value shall be a reference to a constraint for that parameter (i.e., for that substructure in accordance with the definition of the Structured Type). If an ASP definition refers to a parameter specified as being of metatype PDU then in a corresponding constraint, the value for that parameter shall be specified as the name of a PDU constraint, or formal parameter. Use of structured constraints by macro expansion in a constraint shall not be used unless the corresponding ASP definition also references the same Structured Type by macro expansion. Parameterized ASP parameter values in a base constraint shall not be modified or explicitly omitted in a modified constraint */

ASP_ParValues ::= ***$ASP_ParValues*** {*ASP_ParValue*}+ ***$End_ASP_ParValues***

ASP_TypeDefs ::= ***$ASP_TypeDefs*** [*TTCN_ASP_TypeDefs*] [*ASN1_ASP_TypeDefs*] [*ASN1_ASP_TypeDefsByRef*] ***$End_ASP_TypeDefs***

Assignment ::= *DataObjectReference* ":=" *Expression*

/* The LHS of Assignment shall only resolve to: TS_VarIdentifier, TC_-VarIdentifier, reference to the field of a variable or reference to an ASP parameter or PDU field that is to be sent . An expression shall contain no unbound variables . The Expression on the RHS of Assignment shall evaluate to an explicit value of the type of the LHS */

AssignmentList ::= "(" *Assignment* {*Comma Assignment*} ")"

Attach ::= "+" *TreeReference* [*ActualParList*]

/* TreeReference shall not attach itself, either directly or indirectly, at its top level of indentation . The number of the actual parameters shall be the same as the number of the formal parameters. LiteralValue, TS_ParIdentifier, TS_ConstIdentifier, TS_VarIdentifier, TC_VarIdentifier, ConstraintReference dentifier may PCO_Identifier and PCOs CP_Identifier may be passed as actual parameters to an attached tree */

BehaviourDescription ::= ***$BehaviourDescription*** *RootTree* {*LocalTree*} ***$End_BehaviourDescription***

BehaviourLine ::= ***$BehaviourLine*** *LabelId Line Cref VerdictId* [*Comment*] ***$End_BehaviourLine***

Bin ::= ***0*** | ***1***

BitIdentifier ::= *Identifier*

/* BitIdentifier identifies a particular bit within an ASN.1 BIT STRING */

BitNumber ::= *Expression*

/* BitNumber shall evaluate to a non-negative specific INTEGER value */

BitRef ::= *Dot* (*BitIdentifier* | "[" *BitNumber* "]")

BooleanValue ::= ***TRUE*** | ***FALSE***

Bound ::= *Number* | *TS_ParIdentifier* | *TS_ConstIdentifier*

/* Bound shall evaluate to a non-negative INTEGER value or INFINITY */

BoundedFreeText ::= "/*" FreeText "*/"

Bstring ::= " ' " {*Bin* | *Wildcard*} " ' " ***B***

CancelTimer ::= **CANCEL** *[TimerIdentifier]*

CaseIndex ::= ***$CaseIndex*** *TestGroupRef TestCaseId SelExprId Description*

/* Test Cases shall be listed in the order that they exist in the dynamic part. An explicit TestGroupReference shall be provided for the first TestCase of each TestGroup. An explicit TestGroupReference shall be provided for each TestCase that immediately follows a TestGroup */

Char ::= /* A character defined by the relevant character string type. If the CharacterString type includes the character " (double quote), this character shall be represented by a pair of " (double quote) in the denotation of any value */

CharacterString ::= ***NumericString*** | ***PrintableString*** | ***TeletexString*** | ***VideotexString*** | ***VisibleString*** | ***IA5String*** | ***GraphicString*** | ***GeneralString***

CM_Constraints ::= ***$CM_Constraints*** *[TTCN_CM_Constraints] [ASN1_CM_Constraints]* ***$End_CM_Constraints***

CM_Id ::= ***$CM_Id*** *CM_Identifier*

CM_Identifier ::= *Identifier*

CM_ParDcl ::= ***$CM_ParDcl*** *CM_ParId CM_ParType [Comment]* ***$End_CM_ParDcl***

CM_ParDcls ::= ***$CM_ParDcls*** *{CM_ParDcl}* ***$End_CM_ParDcls***

CM_ParId ::= ***$CM_ParId*** *CM_ParIdOrMacro*

CM_ParIdentifier ::= *Identifier*

CM_ParIdOrMacro ::= *CM_ParIdentifier | MacroSymbol*

CM_ParType ::= ***$CM_ParType*** *Type&Attributes*

CM_ParValue ::= ***$CM_ParValue*** *CM_ParId ConsValue [Comment]* ***$End_CM_ParValue***

/* STATIC SEMANTICS - If a CM definition refers to a Structured Type as a substructure of a parameter (i.e., with a parameter name) then the corresponding constraint shall have the same parameter name in the corresponding position in the parameter name name column of the constraint and the value shall be a reference to a constraint for that parameter (i.e. for that substructure in accordance with the definition of the Structured Type). If a CM definition refers to a parameter specified as being of metatype PDU then in a corresponding constraint, the value for that parameter shall be specified as the name of a PDU constraint, or formal parameter. Use of structured constraints by macro expansion in a constraint shall not be used unless the corresponding CM definition also references the same Structured Type by macro expansion. Parameterized CM parameter values in a base constraint shall not be modified or explicitly omitted in a modified constraint */

CM_ParValues ::= ***$CM_ParValues*** *{CM_ParValue}* ***$End_CM_ParValues***

CM_TypeDefs ::= ***$CM_TypeDefs*** *[TTCN_CM_TypeDefs] [ASN1_CM_TypeDefs]* ***$End_CM_TypeDefs***

Colon ::= ":"

Comma ::= ","

Comment ::= ***$Comment*** *[BoundedFreeText]*

Complement ::= **COMPLEMENT** *ValueList*

ComplexDefinitions ::= *[ASP_TypeDefs] PDU_TypeDefs [CM_TypeDefs] [AliasDefs]*

ComponentIdentifier ::= *ASP_ParIdentifier | PDU_FieldIdentifier | CM_ParIdentifier | ElemIdentifier | ASN1_Identifier*

ComponentNumber ::= *Expression*

/* ComponentNumber shall evaluate to a non-negative specific INTEGER value */

ComponentPosition ::= "("*Number*")"

ComponentReference ::= *RecordRef* | *ArrayRef* | *BitRef*

/* RecordRef shall be used to reference ASN.1 SEQUENCE, SET and CHOICE components. It shall not be used to reference components of any other ASN.1 type. RecordRef shall be used to reference ASP parameters, PDU fields and structure elements in the tabular form. ArrayRef shall be used to reference ASN.1 SEQUENCE OF and SET OF components. It shall not be used to reference components of any other ASN.1 type */

Configuration ::= ***$Configuration*** *TCompConfigIdentifier*

ConsId ::= ***$ConsId*** *ConsId&ParList*

ConsId&ParList ::= *ConstraintIdentifier* [*FormalParList*]

ConsRef ::= *ConstraintIdentifier* [*ActualCrefParList*]

ConstraintExpression ::= *Expression*

/* The terms in ConstraintExpression shall only Value, TS_ParIdentifier, TS_ConstIdentifier, FormalParIdentifier and OpCall. ConstraintExpression shall evaluate to an element of the specified type */

ConstraintIdentifier ::= *Identifier*

ConstraintReference ::= *ConsRef* | *FormalParIdentifier*

/*ConsRef shall be present in conjunction with SEND, IMPLICIT SEND and RECEIVE. A ConstraintReference is not needed for ASPs that have no parameters. It shall not be present with any other kind of TTCN statement. FormalParIdentifier shall resolve to a ConsRef */. Value, TS_ParIdentifier, TS_ConstIdentifier, TS_VarIdentifier, TC_VarIdentifier, ConsRef and FormalParIdentifier may be passed as actual parameters to a constraint in a ConstraintReference made from a behaviour description. ConstraintReferences on SEND events shall not include wildcards unless these are explicitly assigned specific values on the SEND event line */

ConstraintsPart ::= ***$ConstraintsPart*** [*TS_TypeConstraints*] [*ASP_Constraints*] [*PDU_Constraints*] [*CM_Constraints*] ***$End_ConstraintsPart***

ConstraintValue&Attributes ::= *ConstraintValue ValueAttributes*

/* ConstraintValue shall fulfil all restrictions defined for the ASP parameter, PDU field or structure element type, including value ranges, value lists, alphabet restrictions and/or length restrictions. Any length specifications defined for the ASP parameter or PDU field type in the Test Suite Type declarations shall not conflict with the length specifications in the ASP or PDU type definition. Neither Test Suite Variables nor Test Case Variables shall be used in constraints, unless passed as actual parameters. In the latter case they shall be bound to a value and shall not be changed */

ConstraintValue&AttributesOrReplace ::= *ConstraintValue&Attributes* | *Replacement* {*Comma Replacement*}

ConstraintValue::= *ConstraintExpression* | *MatchingSymbol* | *ConsRef*

/* LiteralValue, TS_ParIdentifier, TS_ConstIdentifier, TS_VarIdentifier, TC_VarIdentifier, (a different) ConsRef and FormalParIdentifier may be passed as actual parameters in ConsRef */

Construct ::= *GoTo* | *Attach* | *Repeat* | *Return* | *Activate* | *Create* | *No_Defaults*

ConsValue ::= ***$ConsValue*** *ConstraintValue&Attributes*

/* ConsValue shall evaluate to an element of the type specified for the ASP parameter, PDU field or structure element */

CPs_Used ::= ***$CPs_Used*** *[CP_List]*

CP_Dcl ::= ***$CP_Dcl*** *CP_Id [Comment]* ***$End_CP_Dcl***

CP_Dcls ::= ***$Begin_CP_Dcls*** *{CP_Dcl}+ [Comment]* ***$End_CP_Dcls***

CP_Id ::= ***$CP_Id*** *CP_Identifier*

CP_Identifier ::= *Identifier*

CP_List ::= *CP_Identifier {Comma CP_Identifier }*

Create ::= ***CREATE*** "(" *CreateList* ")"

CreateList ::= *CreateTComp {Comma CreateTComp}*

CreateTComp ::= *TCompIdentifier Colon TreeReference [ActualParList]*

Cref ::= ***$Cref*** *[ConstraintReference]*

Cstring ::= " " " *{Char | Wildcard |* "\"} " " "

Dash ::= "-"

DataObjectIdentifier ::= TS_ParIdentifier |TS_ConstIdentifier |TS_VarIdentifier |TC_VarIdentifier |FormalParIdentifier | ASP_Identifier | PDU_Identifier | CM_Identifier

DataObjectReference ::= DataObjectIdentifier {ComponentReference}

/* Identifiers of ASP parameters and PDU fields associated with SEND and RECEIVE shall be used only to reference ASP parameter and PDU field values on the statement line itself. Each ComponentReference shall only reference an ASP parameter, PDU field, structure element or ASN.1 value explicitly declared in the object that immediately precedes in the DataObjectReference */

Declarations ::= *[TS_ConstDcls] [TS_VarDcls] [TC_VarDcls] PCO_Dcls [CP_Dcls] [TimerDcls] [TCompDcls TCompConfigDcls]*

DeclarationsPart ::= ***$DeclarationsPart*** *Definitions Parameterization&Selection Declarations ComplexDefinitions* ***$End_DeclarationsPart***

DeclarationValue ::= *Expression*

/* DeclarationValue shall only contain LiteralValues, TS_ParIdentifiers and TS_ConstIdentifiers. DeclarationValue shall evaluate to an element of its declared type */

Default ::= ***$Begin_Default*** *DefaultId DefaultRef Objective [Comment] BehaviourDescription [Comment]* ***$End_Default***

/* BehaviourDescription shall consist of only one tree (i.e., no local trees). BehaviourDescription shall not use tree attachment except for attaching local trees, i.e. default behaviour trees shall not attach Test Steps). A final verdict shall be assigned to every leaf of a Default tree. If the final verdict results from an OTHERWISE statement in the Default tree the verdict shall be FAIL */

DefaultGroup ::= ***$DefaultGroup*** *DefaultGroupId {DefaultGroup | Default}+* ***$End_DefaultGroup***

DefaultGroupId ::= ***$DefaultGroupId*** *DefaultGroupIdentifier*

DefaultGroupIdentifier ::= *Identifier*

DefaultGroupReference ::= *[SuiteIdentifier* "/"] *{DefaultGroupIdentifier* "/"}

/* There shall be no white space on either side of the "/"s */

DefaultId ::= ***$DefaultId*** *DefaultId&ParList*

DefaultId&ParList ::= *DefaultIdentifier [FormalParList]*

DefaultIdentifier ::= *Identifier*

DefaultIndex ::= ***$Begin_DefaultIndex*** {*DefIndex*} [*Comment*] ***$End_DefaultIndex***

DefaultRef ::= ***$DefaultRef*** *DefaultGroupReference*

DefaultReference ::= *DefaultIdentifier* [*ActualParList*]

DefaultRefList ::= *DefaultReference*{*Comma DefaultReference*}

DefaultsLibrary ::= ***$DefaultsLibrary*** ({*DefaultGroup* | *Default*}+) ***$End_DefaultsLibrary***

DefaultsRef::= ***$DefaultsRef*** [*DefaultRefList*]

DefIndex ::= ***$DefIndex*** *DefaultRef DefaultId Description* ***$End_DefIndex***

/* DefaultId shall not include a formal parameter list. Defaults shall be listed in the order that they exist in the dynamic part. An explicit DefaultGroupReference shall be provided for the first Default of each DefaultGroup. An explicit DefaultGroupReference shall be provided for each Default that immediately follows a DefaultGroup */

Definitions ::= [*TS_TypeDefs*] [*TS_OpDefs*]

DerivationPath ::= {*ConstraintIdentifier Dot*}+

/* If a constraint definition is a modification of an existing constraint, the name of the constraint that is taken as the basis of this modification shall be referenced in the table in the derivation path entry. The first ConstraintIdentifier in DerivationPath shall be a base constraint identifier. Constraints shall be listed in the order in which their modifications to the base constraint are to be applied. There shall be no white space between ConstraintIdentifier and Dot */

DerivPath ::= ***$DerivPath*** [*DerivationPath*]

Description ::= ***$Description*** *BoundedFreeText*

Done ::= "?" ***DONE*** "(" [*TCompIdList*] ")"

Dot ::= "."

Duration ::= ***$Duration*** [*DeclarationValue*]

/* DeclarationValue shall evaluate to a non-zero positive INTEGER */

DynamicPart ::= ***$DynamicPart*** *TestCases* [*TestStepLibrary*] [*DefaultsLibrary*] ***$End_DynamicPart***

ElemDcl ::= ***$ElemDcl*** *ElemId ElemType* [*Comment*] ***$End_ElemDcl***

ElemDcls ::= ***$ElemDcls*** {*ElemDcl*}+ ***$End_ElemDcls***

ElemId ::= ***$ElemId*** *ElemId&FullId*

ElemId&FullId ::= *ElemIdentifier* [*FullIdentifier*]

ElemIdentifier ::= *Identifier*

ElemType ::= ***$ElemType*** *Type&Attributes*

/* There shall be no recursive references (neither directly nor indirectly) in Type&Attributes. A structure element Type shall be a PredefinedType, TS_TypeIdentifier, PDU_Identifier, or PDU */

ElemValue ::= ***$ElemValue*** *ElemId ConsValue* [*Comment*] ***$End_ElemValue***

/* Parameterized Element values in a base constraint shall not be modified or explicitly omitted in a modified constraint */

ElemValues ::= ***$ElemValues*** {*ElemValue*}+ ***$End_ElemValues***

Event ::= *Send* | *Receive* | *Otherwise* | *Timeout* | *Done*

/* A Receive, Otherwise or Timeout event shall only be followed by other Receive, Otherwise and Timeout events through the remainder of the set of alternatives in a fully expanded tree. As a consequence, Default trees will contain only Receive, Otherwise and Timeout events on the first level of alternatives */

ExpandedId ::= ***$ExpandedId*** *Expansion*

Expansion ::= *ASP_Identifier | PDU_Identifier*

Expression ::= *SimpleExpression [RelOp SimpleExpression]*

/* If both SimpleExpressions and the RelOp exist then the SimpleExpressions shall evaluate to specific values of compatible type. If RelOp is "<" | ">" | ">=" | "<=" then each SimpleExpression shall evaluate to a specific INTEGER value. ASN.1 Named Values shall not be used within arithmetic expressions as operands of operations */

ExtendedAlphaNum ::= /* REFERENCE - A character from any character set defined in ISO/IEC 10646 */

Factor ::= *[UnaryOp] Primary*

/* The Primary shall resolve to a specific value. If UnaryOp exists and is "NOT" then Primary shall resolve to type BOOLEAN; if the UnaryOp is "+" or "-" then Primary shall resolve to type INTEGER */

Fail ::= ***FAIL*** | ***F*** | "(" ***FAIL*** ")" | "(" ***F*** ")"

FormalPar&Type ::= *FormalParIdentifier {Comma FormalParIdentifier} Colon FormalParType*

FormalParIdentifier ::= *Identifier*

FormalParList ::= "(" *FormalPar&Type {SemiColon FormalPar&Type}* ")"

FormalParType ::= *Type | PCO_TypeIdentifier* | ***PDU*** | ***CP***

/* If a formal parameter of a Test Step is type PDU then specific fields in the PDU shall not be referenced in the Test Step behaviour tree */

FreeText ::= *{ExtendedAlphaNum}*

/* Free Text shall not contain the string "*/" unless preceded by backslash ("\") */

FullIdentifier ::= "(" *BoundedFreeText* ")"

/* Some TTCN objects allow names, as given in the appropriate protocol standard to be abbreviated. If an abbreviation is used then FullIdentifier shall be given in the declaration of the object */

GoTo ::= ("->" | ***GOTO***) *Label*

/* The label column shall contain labels referenced from the GoTo. Label shall be associated with the first of a set of alternatives, one of which is an ancestor node of the point from which the GoTo is to be made. GoTo shall be used only for jumps within one tree, i.e., within a Test Case root tree, a Test Step tree a Default tree and a local tree; and thus, each label used in a GoTo construct shall be found within the tree in which the GoTo is used.There shall be no ACTIVATE operation as an ancestor node of the GoTo construct on the branch of the tree between the Label and the GoTo. No GoTo shall be made to the first level of alternatives of local trees, Test Steps or Defaults */

Header ::= ***$Header*** *TreeHeader*

Hex ::= *Num* | ***A*** | ***B*** | ***C*** | ***D*** | ***E*** | ***F***

Hstring ::= " ' " *{Hex | Wildcard}* " ' " ***H***

Identifier ::= *Alpha{AlphaNum | Underscore}*

/* All Identifiers referenced in a TTCN test suite shall be explicitly declared in the test suite, explicitly declared in an ASN.1 type definition referenced by the test suite or be a TTCN predefined identifier */

ImplicitSend ::= "<" ***IUT*** "!" (*ASP_Identifier* | *PDU_Identifier*) ">"

/* ImplicitSend shall not be used unless the test method being used is one of the Remote Test Methods */

Inconclusive ::= ***INCONC*** | ***I*** | "(" ***INCONC*** ")" | "(" ***I*** ")"

Indentation ::= "[" *Number* "]"

/* Statements in the first level of alternatives in a behaviour description shall have the indentation value zero. Statements having a predecessor shall have the indentation value of the predecessor plus one as their indentation value */

IntegerRange ::= "(" *LowerTypeBound To UpperTypeBound* ")"

/* LowerTypeBound shall be less than UpperTypeBound */

Label ::= *Identifier*

LabelId ::= ***$LabelId*** [Label]

LengthAttribute ::= *SingleLength* | *RangeLength*

LengthRestriction ::= *SingleTypeLength* | *RangeTypeLength*

/* LengthRestriction shall be provided only when the base type is a string type (i.e., BITSTRING, HEXSTRING, OCTETSTRING or CharacterString) or derived from a string type */

Line ::= ***$Line*** *Indentation StatementLine*

LiteralValue ::= *Number* | *BooleanValue* | *Bstring* | *Hstring* | *Ostring* | *Cstring* | *R_Value*

LocalTree ::= *Header* {*BehaviourLine*}+

LowerAlpha ::= ***a*** | ***b*** | ***c*** | ***d*** | ***e*** | ***f*** | ***g*** | ***h*** | ***i*** | ***j*** | ***k*** | ***l*** | ***m*** | ***n*** | ***o*** | ***p*** | ***q*** | ***r*** | ***s*** | ***t*** | ***u*** | ***v*** | ***w*** | ***x*** | ***y*** | ***z***

LowerBound ::= *Bound*

LowerRangeBound ::= *ConstraintExpression* | *Minus* ***INFINITY***

/* ConstraintExpression shall evaluate to a specific INTEGER value */

LowerTypeBound ::= [*Minus*] *Number* | *Minus* ***INFINITY***

LowerValueBound ::= *ValueBound*

MacroSymbol ::= "<-"

MatchingSymbol ::= *Complement* | *Omit* | *AnyValue* | *AnyOrOmit* | *ValueList* | *ValueRange* | *SuperSet* | *SubSet* | *Permutation*

/* No matching symbol is considered to be a specific value */

Minus ::= "-"

ModuleIdentifier ::= <u>*ModuleIdentifier*</u>

/* Where ModuleIdentifier is a non-terminal defined in ISO/IEC 8824. ModuleIdentifier shall be unique within the domain of interest */

MultiplyOp ::= "*" | "/" | ***MOD*** | ***AND***

/* Operands of the "*", "/" and MOD operators shall be of type INTEGER (i.e., TTCN or ASN.1 predefined) or derivations of INTEGER (i.e., subrange). Operands of the AND operator shall be of type BOOLEAN (TTCN or ASN.1 predefined) or derivatives of BOOLEAN */

MuxValue ::= *TS_ParIdentifier*

No_Defaults ::= ***NO_DEFAULTS***

NonZeroNum ::= ***1*** | ***2*** | ***3*** | ***4*** | ***5*** | ***6*** | ***7*** | ***8*** | ***9***

Num ::= ***0*** | *NonZeroNum*

Number ::= (*NonZeroNum* {*Num*}) | ***0***

NumOf_CPs ::= ***$NumOf_CPs*** *Num_CPs*

NumOf_PCOs ::= ***$NumOf_PCOs*** *Num_PCOs*

Num_CPs ::= *Number*

Num_PCOs ::= *Number*

Objective ::= ***$Objective*** *BoundedFreeText*

Oct ::= *Hex Hex*

Omit ::= *Dash* | ***OMIT***

/*In ASN.1 constraints Omit shall be used only for ASP parameters or PDU fields that are declared OPTIONAL or DEFAULT */

OpCall ::= *OpIdentifier* (*ActualParList* | "(" ")")

/* See static semantics on production ActualParList */

OpIdentifier ::= *TS_OpIdentifier* | *PredefinedOpIdentifier*

Ostring ::= " ' " {*Oct* | *Wildcard*} " ' " ***O***

Otherwise ::= [*PCO_Identifier* | *FormalParIdentifier*] "?" ***OTHERWISE***

/* PCO_Identifier or FormalParIdentifier shall be present if the test suite uses more than one PCO. FormalParIdentifier shall only be of PCO type */

Parameterization&Selection ::= [*TS_ParDcls*] [*SelectExprDefs*]

Pass ::= ***PASS*** | ***P*** | "(" ***PASS*** ")" | "(" ***P*** ")"

PCOs_Used ::= ***$PCOs_Used*** [*PCO_List*]

PCO_Dcl ::= ***$PCO_Dcl*** *PCO_Id PCO_TypeId P_Role* [*Comment*] ***$End_PCO_Dcl***

PCO_Dcls ::= ***$Begin_PCO_Dcls*** {*PCO_Dcl*}+ [*Comment*] ***$End_PCO_Dcls***

/* In accordance with ISO/IEC 9646-1 the number of PCOs shall relate to the test method used */

PCO_Id ::= ***$PCO_Id*** *PCO_Identifier*

PCO_Identifier ::= *Identifier*

PCO_List ::= *PCO_Identifier* {*Comma PCO_Identifier* }

PCO_Role ::= ***UT*** | ***LT***

PCO_Type ::= ***$PCO_Type*** [*PCO_TypeIdentifier* ["("*MuxValue*")"]]

/* PCO_TypeIdentifier shall be one of the PCO types used in the PCO declaration proforma. If only a single PCO is defined within a test suite then PCO_TypeIdentifier is optional */

PCO_TypeId ::= ***$PCO_TypeId*** *PCO_TypeIdentifier*

PCO_TypeIdentifier ::= *Identifier*

PDU_Constraints ::= ***$PDU_Constraints*** [*TTCN_PDU_Constraints*] [*ASN1_PDU_Constraints*] ***$End_PDU_Constraints***

PDU_FieldDcl ::= ***$PDU_FieldDcl*** *PDU_FieldId PDU_FieldType* [*Comment*] ***$End_PDU_FieldDcl***

PDU_FieldDcls ::= ***$PDU_FieldDcls*** {*PDU_FieldDcl*}+ ***$End_PDU_FieldDcls***

PDU_FieldId ::= ***$PDU_FieldId*** *PDU_FieldIdOrMacro*

PDU_FieldId&FullId ::= *PDU_FieldIdentifier* [*FullIdentifier*]

PDU_FieldIdentifier ::= *Identifier*

PDU_FieldIdOrMacro ::= *PDU_FieldId*&FullId | *MacroSymbol*

/* The MacroSymbol shall be used only in combination with a reference to a Structured Type */

PDU_FieldType ::= ***$PDU_FieldType*** *Type&Attributes*

/* Type shall be a PredefinedType or TS_TypeIdentifier, PDU_Identifier, or PDU */

PDU_FieldValue ::= ***$PDU_FieldValue*** *PDU_FieldId ConsValue* [*Comment*] ***$End_PDU_FieldValue***

/* The FullIdentifier that is part of PDU_FieldId shall not be used. If a PDU definition refers to a Structured Type as a substructure of a field (i.e., with a field name) then the corresponding constraint shall have the same field name in the corresponding position in the field name name column of the constraint and the value shall be a reference to a constraint for that field (i.e., for that substructure in accordance with the definition of the Structured Type). If a PDU definition refers to a field specified as being of metatype PDU then in a corresponding constraint, the value for that field shall be specified as the name of a PDU constraint, or formal parameter. Use of structured constraints by macro expansion in a constraint shall not be used unless the corresponding PDU definition also references the same Structured Type by macro expansion. Parameterized PDU field values in a base constraint shall not be modified or explicitly omitted in a modified constraint */

PDU_FieldValues ::= ***$PDU_FieldValues*** {*PDU_FieldValue*}+ ***$End_PDU_FieldValues***

PDU_Id ::= ***$PDU_Id*** *PDU_Id&FullId*

PDU_Id&FullId ::= *PDU_Identifier* [*FullIdentifier*]

PDU_Identifier ::= *Identifier*

PDU_TypeDefs ::= ***$PDU_TypeDefs*** [*TTCN_PDU_TypeDefs*] [*ASN1_PDU_TypeDefs*] [*ASN1_PDU_TypeDefsByRef*] ***$End_PDU_TypeDefs***

Permutation ::= ***PERMUTATION*** *ValueList*

/* The Permutation shall be used only inside a value of type SEQUENCE OF. The ValueList shall be of the type specified in the SEQUENCE OF */

PICSref ::= ***$PICSref*** *BoundedFreeText*

PICS_PIXITref ::= ***$PICS_PIXITref*** *BoundedFreeText*

PIXITref ::= ***$PIXITref*** *BoundedFreeText*

PredefinedOpIdentifier ::= ***BIT_TO_INT*** | ***HEX_TO_INT*** | ***INT_TO_BIT*** | ***INT_TO_HEX*** | ***IS_CHOSEN*** | ***IS_PRESENT*** | ***LENGTH_OF*** | ***NUMBER_OF_ELEMENTS***

PredefinedType ::= ***INTEGER*** | ***BOOLEAN*** | ***BITSTRING*** | ***HEXSTRING*** | ***OCTETSTRING*** | ***R_Type*** | *CharacterString*

Primary ::= *Value* | *DataObjectReference* | *OpCall* | *SelectExprIdentifier* | "(" *Expression* ")"

/* SelectExprIdentifier shall only be used within selection expressions */

P_Role ::= ***$PCO_Role*** *PCO_Role*

Qualifier ::= "[" *Expression* "]"

/* Qualifier shall evaluate to a specific BOOLEAN value */

RangeLength ::= "[" *LowerBound To UpperBound* "]"

/* LowerBound shall be less than UpperBound */

RangeTypeLength ::= "[" *LowerTypeBound To UpperTypeBound* "]"

RangeValueLength ::= "[" *LowerValueBound To UpperValueBound* "]"

/* LowerValueBound shall be less than UpperValueBound */

ReadTimer ::= ***READTIMER*** *TimerIdentifier* "(" *DataObjectReference*")"

/* The DataObjectReference shall only resolve to TS_VarIdentifier, TC_-VarIdentifier, reference to the field of a variable or reference to an ASP parameter or PDU field that is to be sent. The DataObjectReference shall resolve to type INTEGER */

Receive ::= [*PCO_Identifier* | *CP_Identifier* | *FormalParIdentifier*] "?" (*ASP_Identifier* | *PDU_Identifier* | *CM_Identifier*)

/* PCO_Identifier or FormalParIdentifier shall be present if the test suite uses more than one PCO. Only CMs may be exchanged on CPs and only ASPs and PDUs may be exchanged on PCOs */

RecordRef ::= *Dot* (*ComponentIdentifier* | *PDU_Identifier* | *StructIdentifier* | *ComponentPosition*)

/* The ComponentIdentifier form of RecordRef shall always be used to reference ASN.1 SEQUENCE, SET and CHOICE components when an identifier is declared for the component. The ComponentIdentifier form of RecordRef shall always be used to reference ASP parameters, PDU fields and structure elements declared in the tabular form. The ComponentPosition form of RecordRef shall always be used to reference ASN.1 SEQUENCE, SET and CHOICE components when an identifier is not declared for the component. StructIdentifier shall not be used if the relevant structure is used as a macro. StructIdentifiers and PDU_Identifiers shall be explicitly included in a RecordRef whenever a parameter, field or element is chained to a PDU or structure and the RecordRef is to identify a component of that PDU or structure. Where a structure is used as a macro expansion, the elements in the structure shall be referred to as if it was expanded into the ASP or PDU referring to it. If a parameter, field or element is defined to be of metatype PDU no reference shall be made to fields of that substructure */

ReferenceList ::= (*ArrayRef* | *ComponentIdentifier* | *ComponentPosition*) {*ComponentReference*}

ReferenceType ::= *TS_TypeIdentifier* | *ASP_Identifier* | *PDU_Identifier* | *CM_Identifier*

/* All types, other than the predefined types, used in a test suite shall be declared in the Test Suite Type definitions or ASP type definitions or PDU type definitions and referenced by name */

RelOp: := "=" | "<" | ">" | "<>" | ">=" | "<="

Repeat ::= ***REPEAT*** *TreeReference* [*ActualParList*] ***UNTIL*** *Qualifier*

/* TreeReference shall not attach itself, either directly or indirectly, at its top level of indentation. The number of the actual parameters shall be the same as the number of the formal parameters. LiteralValue, TS_ParIdentifier, TS_ConstIdentifier, TS_VarIdentifier, TC_VarIdentifier, ConstraintReferenceIdentifier may PCO_Identifier and CP_Identifier PCOs may be passed as actual parameters to the tree in a REPEAT statement */

Replacement ::= (***REPLACE*** *ReferenceList* ***BY*** *ConstraintValue&Attributes*) | (***OMIT*** *ReferenceList*)

/* Replacement shall be used only when DerivPath is specified. Parameteized replaced values in a base constraint shall not be modified or explicitly omitted in a modified constraint */

Restriction ::= *LengthRestriction* | *IntegerRange* | *SimpleValueList*

/* The set of values defined by Restriction shall be a true subset of the values of the base type */

Result ::= ***R***

/* R shall not be used on the LHS of an assignment */

Return ::= ***RETURN***

/* The RETURN construct shall not be used except in Default behaviour trees (including any local trees within Default behaviour tables). RETURN means return from the Default tree. Therefore, it cannot be used anywhere else other than in the Default trees */

RootTree ::= {*BehaviourLine*}+

R_Value ::= ***pass*** | ***fail*** | ***inconc*** | ***none***

SelectExpr ::= ***$SelectExpr*** *SelectionExpression*

SelectExprDef::= ***$SelectExprDef*** *SelectExprId SelectExpr* [*Comment*] ***$End_SelectExprDef***

SelectExprDefs ::= ***$Begin_SelectExprDefs*** {*SelectExprDef*}+ [*Comment*] ***$End_SelectExprDefs***

SelectExprId ::= ***$SelectExprId*** *SelectExprIdentifier*

SelectExprIdentifier ::= *Identifier*

SelectionExpression ::= *Expression*

/* SelectionExpression shall only contain LiteralValues, TS_ParIdentifiers, TS_ConstIdentifiers and SelectExprIdentifiers. SelectionExpression shall evaluate to a specific BOOLEAN value. Expression shall not recursively refer (neither directly nor indirectly) to the SelExprIdentifier being defined by that Expression */

SelExprId ::= ***$SelectExprId*** [*SelectExprIdentifier*]

SemiColon ::= ";"

Send ::= [*PCO_Identifier* | *CP_Identifier* | *FormalParIdentifier*] "!" (*ASP_Identifier* | *PDU_Identifier* | *CM_Identifier*)

/* PCO_Identifier or FormalParIdentifier shall be present if the test suite uses more than one PCO. FormalParIdentifier shall resolve to a PCO_-Identifier. This must be present if the test suite uses more than one PCO. Only CMs may be exchanged on CPs and only ASPs and PDUs may be exchanged on PCOs */

SimpleExpression ::= *Term* {*AddOp Term*}

/* Each Term shall resolve to a specific value. If more than one Term exists and if AddOp is "OR" then the Terms shall resolve to type BOOLEAN; if AddOp is "+" or "-" then the Terms shall resolve to type INTEGER */

SimpleTypeDef ::= ***$SimpleTypeDef*** *SimpleTypeId SimpleTypeDefinition* [*Comment*] ***$End_SimpleTypeDef***

SimpleTypeDefinition ::= ***$SimpleTypeDefinition*** *Type&Restriction*

SimpleTypeDefs ::= ***$Begin_SimpleTypeDefs*** {*SimpleTypeDef*}+ [*Comment*] ***$End_SimpleTypeDefs***

SimpleTypeId ::= ***$SimpleTypeId*** *SimpleTypeIdentifier*

SimpleTypeIdentifier ::= *Identifier*

SimpleValueList::= "(" [*Minus*] *LiteralValue* {*Comma* [*Minus*] *LiteralValue*} ")"

/* If Minus is used in SimpleValueList then LiteralValue shall be a number. The LiteralValues shall be of the base type and shall be a true subset of the values defined by the base type */

SingleLength ::= "[" *Bound* "]"

SingleTypeLength ::= "["*Number* "]"

SingleValueLength ::= "[" *ValueBound* "]"

StandardsRef ::= ***$StandardsRef*** *BoundedFreeText*

StartTimer ::= ***START*** *TimerIdentifier* ["(" *TimerValue* ")"]

StatementLine ::= (*Event* [*Qualifier*] [*AssignmentList*] [*TimerOps*]) | (*Qualifier* [*AssignmentList*] [*TimerOps*]) | (*AssignmentList* [*TimerOps*]) | *TimerOps* |*Construct* | *ImplicitSend*

StepIndex ::= ***$StepIndex*** *TestStepRef TestStepId Description* ***$End_StepIndex***

/* TestStepId shall not include a formal parameter list. Test Steps shall be listed in the order that they exist in the dynamic part. An explicit TestStepGroupReference shall be provided for the first TestStep of each TestStepGroup. An explicit TestSteGroupReference shall be provided for each TestStep that immediately follows a TestStepGroup */

StructId ::= ***$StructId*** *StructId&FullId*

StructId&FullId ::= *StructIdentifier* [*FullIdentifier*]

StructIdentifier ::= *Identifier*

StructTypeConstraint ::= ***$Begin_StructTypeConstraint*** *ConsId StructId DerivPath* [*Comment*] *ElemValues* [*Comment*] ***$End_StructTypeConstraint***

/* The FullIdentifier that is part of Struct_Id shall not be used. A modified constraint shall have the same parameter list as its base constraint. In particular, there shall be no parameters omitted from or added to this list */

StructTypeConstraints ::= ***$StructTypeConstraints*** {*StructTypeConstraint*}+ ***$End_StructTypeConstraints***

StructTypeDef ::= ***$Begin_StructTypeDef*** *StructId* [*Comment*] *ElemDcls* [*Comment*] ***$End_StructTypeDef***

StructTypeDefs ::= ***$StructTypeDefs*** {*StructTypeDef*}+ ***$End_StructTypeDefs***

Structure&Objective ::= ***$Structure&Objective*** *TestGroupRef SelExprId Objective* ***$End_Structure&Objective***

Structure&Objectives ::= ***$Structure&Objectives*** {*Structure&Objective*} ***$End_Structure&Objectives***

SubSet ::= ***SUBSET*** "(" *ConstraintValue&Attributes* ")"

/* The argument to SubSet, i.e., ConstraintValue&Attributes, shall be of type SET OF */

Suite ::= ***$Suite*** *SuiteId SuiteOverviewPart DeclarationsPart ConstraintsPart DynamicPart* ***$End_Suite***

SuiteId ::= ***$SuiteId*** *SuiteIdentifier*

SuiteIdentifier ::= *Identifier*

SuiteOverviewPart ::= ***$SuiteOverviewPart*** *SuiteStructure TestCaseIndex* [*TestStepIndex*] [*DefaultIndex*] ***$End_SuiteOverviewPart***

SuiteStructure ::= ***$Begin_SuiteStructure*** *SuiteId StandardsRef PICSref PIXITref TestMethods* [*Comment*] *Structure&Objectives* [*Comment*] ***$End_SuiteStructure***

SuperSet ::= ***SUPERSET*** "(" *ConstraintValue&Attributes* ")"

/* The argument to SuperSet, i.e., ConstraintValue&Attributes, shall be of type SET OF */

TCompConfigDcl ::= ***$Begin_TCompConfigDcl*** *TCompConfigId* [*Comment*] {*TCompConfigInfo*}+ [*Comment*] ***$End_TCompConfigDcl***

TCompConfigDcls ::= ***$TCompConfigDcls*** {*TCompConfigDcl*}+ ***$End_TCompConfigDcls***

TCompConfigId ::= ***$TCompConfigId*** *TCompConfigIdentifier*

TCompConfigIdentifier ::= *Identifier*

TCompConfigInfo ::= ***$TCompConfigInfo*** *TCompUsed PCOs_Used CPs_Used* [*Comment*]

TCompDcl ::= ***$TCompDcl*** *TCompId TComponentRole NumOf_PCOs NumOf_CPs* [*Comment*] ***$End_TCompDcl***

TCompDcls ::= ***$Begin_TCompDcls*** {*TCompDcl*}+ [*Comment*] ***$End_TCompDcls***

TCompId ::= ***$TCompId*** *TCompIdentifier*

TCompIdentifier::= *Identifier*

TCompIdList ::= *TCompIdentifier* | *{Comma TCompIdentifier}*

TComponentRole ::= ***$TComponentRole*** *TCompRole*

TCompRole ::= ***MTC*** | ***PTC***

TCompUsed ::= ***$TCompUsed*** *TCompIdentifier*

TC_VarDcl ::= ***$TC_VarDcl*** *TC_VarId TC_VarType TC_VarValue* [*Comment*] ***$End_TC_VarDcl***

TC_VarDcls ::= ***$Begin_TC_VarDcls*** {*TC_VarDcl*}+ [*Comment*] ***$End_TC_VarDcls***

TC_VarId ::= ***$TC_VarId*** *TC_VarIdentifier*

TC_VarIdentifier ::= *Identifier*

TC_VarType ::= ***$TC_VarType*** *Type*

/* Type shall be a PredefinedType, TS_TypeIdentifier or a PDU_Identifier */

TC_VarValue ::= ***$TC_VarValue*** [*DeclarationValue*]

Term ::= *Factor* {*MultiplyOp Factor*}

/* Each Factor shall resolve to a specific value. If more than one Factor exists and if MultiplyOp is 'AND' then the Factors shall resolve to type BOOLEAN; if MultiplyOp is '*' or '/' then the Factors shall resolve to type INTEGER */

TestCase ::= ***$Begin_TestCase*** *TestCaseId TestGroupRef TestPurpose* [*Configuration*] *DefaultsRef* [*Comment*] *BehaviourDescription* [*Comment*] ***$End_TestCase***

TestCaseId ::= ***$TestCaseId*** *TestCaseIdentifier*

TestCaseIdentifier ::= *Identifier*

TestCaseIndex ::= ***$Begin_TestCaseIndex*** {*CaseIndex*}+ [*Comment*] ***$End_TestCaseIndex***

TestCases ::= ***$TestCases*** ({*TestGroup* | *TestCase*}+) ***$End_TestCases***

TestGroup ::= ***$TestGroup*** *TestGroupId* {*TestGroup* | *TestCase*}+ ***$End_TestGroup***

TestGroupId ::= ***$TestGroupId*** *TestGroupIdentifier*

TestGroupIdentifier ::= *Identifier*

TestGroupRef ::= ***$TestGroupRef*** *TestGroupReference*

TestGroupReference ::= [SuiteIdentifier "/"] {*TestGroupIdentifier* "/"}

/* There shall be no white space on either side of the "/"s */

TestMethods ::= ***$TestMethods*** *BoundedFreeText*

TestPurpose ::= ***$TestPurpose*** *BoundedFreeText*

TestStep ::= ***$Begin_TestStep*** *TestStepId TestStepRef Objective DefaultsRef* [*Comment*] *BehaviourDescription* [*Comment*] ***$End_TestStep***

TestStepGroup ::= ***$TestStepGroup*** *TestStepGroupId* {*TestStepGroup* | *TestStep*}+ ***$End_TestStepGroup***

TestStepGroupId ::= ***$TestStepGroupId*** *TestStepGroupIdentifier*

TestStepGroupIdentifier ::= *Identifier*

TestStepGroupReference ::= [*SuiteIdentifier* "/"] {*TestStepGroupIdentifier* "/"}
/* There shall be no white space on either side of the "/"s */

TestStepId ::= ***$TestStepId*** *TestStepId&ParList*

TestStepId&ParList ::= *TestStepIdentifier* [*FormalParList*]

TestStepIdentifier ::= *Identifier*

TestStepIndex ::= ***$Begin_TestStepIndex*** {*StepIndex*} [*Comment*] ***$End_TestStepIndex***

TestStepLibrary ::= ***$TestStepLibrary*** ({*TestStepGroup* | *TestStep*}+) ***$End_TestStepLibrary***

TestStepRef ::= ***$TestStepRef*** *TestStepGroupReference*

Timeout ::= "?" ***TIMEOUT*** [*TimerIdentifier*]

TimerDcl ::= ***$TimerDcl*** *TimerId Duration Unit* [*Comment*] ***$End_TimerDcl***

TimerDcls ::= ***$Begin_TimerDcls*** {*TimerDcl*}+ [*Comment*] ***$End_TimerDcls***

TimerId ::= ***$TimerId*** *TimerIdentifier*

TimerIdentifier ::= *Identifier*

TimerOp ::= *StartTimer* | *CancelTimer* | *ReadTimer*

TimerOps ::= *TimerOp* {*Comma TimerOp*}

TimerValue ::= *Expression*
/* Timer value shall evaluate to a non-zero positive INTEGER */

TimeUnit ::= ***ps*** | ***ns*** | ***us*** | ***ms*** | ***s*** | ***min***
/* If a timer is derived from the PICS/PIXIT then the timer declaration shall specify the same units as the PICS/PIXIT entry */

To ::= ***TO*** | ".."

TreeHeader ::= *TreeIdentifier* [*FormalParList*]

TreeIdentifier ::= *Identifier*

TreeReference ::= *TestStepIdentifier* | *TreeIdentifier*
/* TreeIdentifier shall be the name of one of the trees in the current behaviour description, i.e., local trees are not accessible outside the behaviour description in which they are specified */

TS_ConstDcl ::= ***$TS_ConstDcl*** *TS_ConstId TS_ConstType TS_ConstValue* [*Comment*] ***$End_TS_ConstDcl***

TS_ConstDcls ::= ***$Begin_TS_ConstDcls*** {*TS_ConstDcl*}+ [*Comment*] ***$End_TS_ConstDcls***

TS_ConstId ::= ***$TS_ConstId*** *TS_ConstIdentifier*

TS_ConstIdentifier ::= *Identifier*

TS_ConstType ::= ***$TS_ConstType*** *Type*
/* Type shall be a PredefinedType, TS_TypeIdentifier or a PDU_Identifier */

TS_ConstValue ::= ***$TS_ConstValue*** *DeclarationValue*

TS_OpDef ::= ***$Begin_TS_OpDef*** *TS_OpId TS_OpResult* [*Comment*] *TS_OpDescription* [*Comment*] ***$End_TS_OpDef***

TS_OpDefs ::= ***$TS_OpDefs*** {*TS_OpDef*}+ ***$End_TS_OpDefs***

TS_OpDescription ::= *$TS_OpDescription BoundedFreeText*

TS_OpId ::= ***$TS_OpId*** *TS_OpId&ParList*

TS_OpId&ParList ::= *TS_OpIdentifier* [*FormalParList*]
/* A Test Suite Operation formal parameter Type shall be a PredefinedType, TS_TypeIdentifier or PDU_Identifier.*/

TS_OpIdentifier ::= *Identifier*

TS_OpResult ::= ***$TS_OpResult*** *Type*
/* Type shall be a PredefinedType, TS_TypeIdentifier or PDU_Identifier */

TS_ParDcl ::= ***$TS_ParDcl*** *TS_ParId TS_ParType PICS_PIXITref* [*Comment*] ***$End_TS_ParDcl***

TS_ParDcls ::= ***$Begin_TS_ParDcls*** {*TS_ParDcl*}+ [*Comment*] ***$End_TS_ParDcls***

TS_ParId ::= ***$TS_ParId*** *TS_ParIdentifier*

TS_ParIdentifier ::= *Identifier*

TS_ParType ::= ***$TS_ParType*** *Type*
/* Type shall be a PredefinedType, TS_TypeIdentifier or a PDU_Identifier */

TS_TypeConstraints ::= ***$TS_TypeConstraints*** [*StructTypeConstraints*] [*ASN1_TypeConstraints*] ***$End_TS_TypeConstraints***

TS_TypeDefs ::= ***$TS_TypeDefs*** [*SimpleTypeDefs*] [*StructTypeDefs*] [*ASN1_TypeDefs*] [*ASN1_TypeRefs*] ***$End_TS_TypeDefs***

TS_TypeIdentifier ::= *SimpleTypeIdentifier* | *StructIdentifier* | *ASN1_TypeIdentifier*

TS_VarDcl ::= ***$TS_VarDcl TS_VarId*** *TS_VarType TS_VarValue* [*Comment*] ***$End_TS_VarDcl***

TS_VarDcls ::= ***$Begin_TS_VarDcls*** {*TS_VarDcl*}+ [*Comment*] ***$End_TS_VarDcls***

TS_VarId ::= ***$TS_VarId*** *TS_VarIdentifier*

TS_VarIdentifier ::= *Identifier*

TS_VarType ::= ***$TS_VarType*** *Type*
/* Type shall be a PredefinedType, TS_TypeIdentifier or a PDU_-Identifier */

TS_VarValue ::= ***$TS_VarValue*** [*DeclarationValue*]

TTCN_ASP_Constraint ::= ***$Begin_TTCN_ASP_Constraint*** *ConsId ASP_Id DerivPath* [*Comment*] *ASP_ParValues* [Comment] ***$End_TTCN_ASP_Constraint***
/* The FullIdentifier that is part of ASP_Id shall not be used. If an ASP is substructured, then the constraints for ASPs of that type shall have the same structure. For every ASP type definition at least one base constraint shall be specified. A modified constraint shall have the same parameter list as its base constraint. In particular, there shall be no parameters omitted from or added to this list */

TTCN_ASP_Constraints ::= ***$TTCN_ASP_Constraints*** {*TTCN_ASP_Constraint*}+ ***$End_TTCN_ASP_Constraints***

TTCN_ASP_TypeDef ::= ***$Begin_TTCN_ASP_TypeDef*** *ASP_Id PCO_Type* [*Comment*] *ASP_ParDcls* [*Comment*] ***$End_TTCN_ASP_TypeDef***

TTCN_ASP_TypeDefs ::= ***$TTCN_ASP_TypeDefs*** {*TTCN_ASP_TypeDef*}+ ***$End_TTCN_ASP_TypeDefs***

TTCN_CM_Constraint ::= ***$Begin_TTCN_CM_Constraint*** *ConsId&Parlist CM_Id* [*DerivationPath*] [*Comment*] *CM_ParValues* [*Comment*] ***$End_TTCN_CM_Constraint***
/* STATIC SEMANTICS - If a CM is substructured, then the constraints for CMs of that type shall have the same structure. For every CM type definition at least one base constraint shall be specified. A modified constraint shall have the same parameter list as its base constraint. In particular, there shall be no parameters omitted from or added to this list */

TTCN_CM_Constraints ::= ***$TTCN_CM_Constraints*** {*TTCN_CM_Constraint*}+ ***$End_TTCN_CM_Constraints***

TTCN_CM_TypeDef ::= ***$Begin_TTCN_CM_TypeDef*** *CM_Id* [*Comment*] *CM_ParDcls* [*Comment*] ***$End_TTCN_CM_TypeDef***

TTCN_CM_TypeDefs ::= ***$TTCN_CM_TypeDefs*** {*TTCN_CM_TypeDef*}+ ***$End_TTCN_CM_TypeDefs***

TTCN_PDU_Constraint ::= ***$Begin_TTCN_PDU_Constraint*** *ConsId PDU_Id DerivPath* [*Comment*] PDU_FieldValues [*Comment*] ***$End_TTCN_PDU_Constraint***

/* The FullIdentifier that is part of PDU_Id shall not be used. If a PDU is substructured, then the constraints for PDUs of that type shall have the same structure. For every PDU type definition at least one base constraint shall be specified. A modified constraint shall have the same parameter list as its base constraint. In particular, there shall be no parameters omitted from or added to this list */

TTCN_PDU_Constraints ::= ***$TTCN_PDU_Constraints*** {*TTCN_PDU_Constraint*}+ ***$End_TTCN_PDU_Constraints***

TTCN_PDU_TypeDef ::= ***$Begin_TTCN_PDU_TypeDef*** *PDU_Id PCO_Type* [*Comment*] *PDU_FieldDcls* [*Comment*] ***$End_TTCN_PDU_TypeDef***

/* If a PDU is sent or received only embedded in ASPs within the whole test suite, then PCO_TypeIdentifier (in PCO_Type) is optional */

TTCN_PDU_TypeDefs ::= ***$TTCN_PDU_TypeDefs*** {*TTCN_PDU_TypeDef*}+ ***$End_TTCN_PDU_TypeDefs***

Type ::= *PredefinedType* | *ReferenceType*

Type&Attributes::= (*Type* [*LengthAttribute*]) | ***PDU***

/* The set of values defined by LengthAttribute shall be a true subset of the values of the base type.LengthAttribute shall be provided only when the base type is a string type (i.e., BITSTRING, HEXSTRING, OCTETSTRING or CharacterString) or derived from a string type */

Type&Restriction ::= *Type* [*Restriction*]

TypeReference ::= *typereference*

/* Where typereference is defined in ISO/IEC 8824:1990, subclause 8.2 */

UnaryOp ::= "+" | "-" | ***NOT***

/* Operands of the "+", "-" operators shall be of type INTEGER (i.e., TTCN or ASN.1 predefined) or derivations of INTEGER (i.e., subrange). Operands of the NOT operator shall be of type BOOLEAN (TTCN or ASN.1 predefined) or derivatives of BOOLEAN */

Underscore ::= "_"

Unit ::= ***$Unit*** *TimeUnit*

UpperAlpha ::= ***A*** | ***B*** | ***C*** | ***D*** | ***E*** | ***F*** | ***G*** | ***H*** | ***I*** | ***J*** | ***K*** | ***L*** | ***M*** | ***N*** | ***O*** | ***P*** | ***Q*** | ***R*** | ***S*** | ***T*** | ***U*** | ***V*** | ***W*** | ***X*** | ***Y*** | ***Z***

UpperBound ::= *Bound* | ***INFINITY***

UpperRangeBound ::= *ConstraintExpression* | ***INFINITY***

/* ConstraintExpression shall evaluate to a specific INTEGER value */

UpperTypeBound ::= [*Minus*] *Number* | ***INFINITY***

UpperValueBound ::= *ValueBound* | ***INFINITY***

ValRange ::= *LowerRangeBound To UpperRangeBound*

/* LowerRangeBound shall be less than UpperRangeBound */

Value ::=	*LiteralValue* | *ASN1_Value*
	/*Where ASN1_Value is Value as defined in ISO/IEC 8824. In ISO/IEC 8824 the production DefinedValue is defined as: DefinedValue ::= Externalvaluereference | valuereference. For the purposes of TTCN this production is redefined to be: DefinedValue ::= ConstraintValue&Attributes. Note that this means that external references are not allowed in TTCN. ASN.1 Named Values shall not be used within arithmetic expressions as operands of operations */
ValueAttributes ::=	[*ValueLength*] [***IF_PRESENT***]
	/* In ASN.1 constraints IF_PRESENT shall be used only for ASP parameters or PDU fields that are declared OPTIONAL or DEFAULT */
ValueBound ::=	*Number* | *TS_ParIdentifier* | *TS_ConstIdentifier* | *FormalParIdentifier*
	/* ValueBound shall evaluate to a specific non-negative INTEGER value */
ValueLength ::=	*SingleValueLength* | *RangeValueLength*
	/* ValueLength shall be used only for ASP parameters, PDU fields or structure element that are declared as BITSTRING, HEXSTRING, OCTETSTRING, CharacterString, SEQUENCE OF or SET OF. ValueLength shall be used only in combination with the following mechanisms: Specific value, Complement, Omit, AnyValue, AnyOrOmit, AnyOrNone and Permutation. The set of values defined by ValueLength shall be a true subset of the values allowed by the ASP parameter's, PDU field's or structure element's declared type */
ValueList ::=	"("*ConstraintValue&Attributes* {*Comma ConstraintValue&Attributes* }")"
	/* Each ConstraintValue&Attributes shall be of the type declared for the ASP parameter, PDU field, or structure element in which the ValueList is used */
ValueRange ::=	"(" *ValRange* ")"
	/* ValueRange shall be used only on ASP parameter, PDU field, or structure element of type INTEGER. The set of values defined by ValueRange shall be a true subset of the values allowed by the ASP parameter's, PDU field's or structure element's declared type */
Verdict ::=	*Pass* | *Fail* | *Inconclusive* | *Result*
	/* Verdict shall not occur corresponding to entries in the behaviour tree which are any of the following: empty, an ATTACH construct, a REPEAT construct, a GOTO construct, an IMPLICIT SEND */
VerdictId ::=	***$VerdictId*** [*Verdict*]
Wildcard ::=	*AnyOne* | *AnyOrNone*

ANNEX B. AN OVERVIEW OF TEST TOOLS

This annex provides information on existing test tools related to CTMF and TTCN that the authors are aware of. For each test tool listed, its name, category and address information are given. Categories are TTCN.GR editor, TTCN.GR/MP converter, TTCN.MP compiler, administration of testing documents (PICS, PIXIT), test case generation, test case validation, results analysis, test execution for protocol area, etc. Address information includes the postal address, contact person, phone number, fax number and e-mail address of the tool suppliers. For test tools for which the authors could not procure this kind of information at least a reference is included.

The authors are certainly not aware of every relevant tool available, though they tried hard to identify as many as they could. Most of the material is based on information given some time or other by the tool providers, but the wording used below should not be considered as authorized by the providers of the tools. Readers looking for authorized and up to date information on a tool should contact its providers. The authors cannot accept any responsibility as to the accurateness of the information they give. The authors are not expressing any judgement on the tools mentioned; in particular, the extent of the information they give is merely a consequence of the number of tool components and the details they could find out, and does not imply any preference. Any name mentioned may be a trademark, even if not marked as such. The reservations made with respect to the authors hold equally with respect to their employer. The authors plan to provide a TTCN and CTMF tool information service, by collecting tool informations e-mailed to them (up to 2000 characters per provider) and distributing them in electronic form. This will, hopefully, suffice to do justice to providers who happened not to be considered in this appendix.

TABLE B.1. Test tools

TEST TOOL	CATEGORY
CICA	Ferry based test system, PICS-, PIXIT editor, TTCN-GR editor, TTCN-GR/ MP converter, TTCN-MP/C compiler
Reference [PTSV 93], pp. 325-326	
Conformance Kit	
Reference [PTSV 93], p. 332	

TABLE B.1 continued(1). Test tools

TEST TOOL	CATEGORY
CTTS	TTCN.MP Compiler, TTCNMP/ TTCN.GR translator, test execution with interactive user interface, test case generation, test execution, test result analysing system for layers 2-4
PICS-CED	administration and processing of PICS and PIXIT proformas
TCDM	test campaign document manager system, production of test reports
Siemens Nixdorf Informationssysteme AG, Dept. MR PD 454 O. Lang, Otto-Hahn-Ring 6 D-81730 München Germany	Tel: 49 89 636 47611 Fax: +49 89 636 40044
Explosions-Toolset	TTCN.MP test sequence generation tool based on Estelle specifications
Olaf Henniger Institut für Automatisierungstechnik Otto-von-Guericke-Universität Magdeburg POBox 4120 D-39016 Magdeburg Germany	Tel: +49 391 5529 2732 Fax: +49 391 5592 2229 E-Mail: henniger@infaut.e-technik.tu-magdeburg.de
FSMdiags	diagnosis of implementations modeled by finite state machines
FCA	fault coverage analysis for a given test suite and incremental test generation based on a given specification in the form of completely and partially specified finite state machine
TANGO	trace analysis of Estelle specifications
TESTL	Estelle based incremental test design environment, validation and generation of TTCN test suites
TETRA	analysis tool for traces and test cases written in (or translated to) LOTOS
TSE	evaluation of test coverage
Gregor v. Bochmann D.d'IRO, University of Montreal 6128, succ. Centre-Ville Montreal (Quebec) H3C 3J7, Canada	Tel: (514) 347 7484 Fax: (514) 343 5834 E-mail: bochmann@iro.umontreal.ca
GENEPX400-PC	development, integration and acceptance testing tool
Reference [PTS 90], p. 369	

TABLE B.1 continued(2). Test tools

TEST TOOL	CATEGORY
GeneTest	automatic generation of TTCN test cases based on Petri net protocol specifications
Bernd Baumgarten GMD/I2 Rheinstr. 75 D-64295 Darmstadt Germany	Tel: +49 6151 869 263 Fax: +49 6151 869 224 E-mail: baumgart@darmstadt.gmd.de
IDACOM PT 500	X.25, ISDN test tool
Reference [PTSV 93], p. 330	
INTEGRATOR II	protocol analysis and test execution tool for ISO 8473, ISO 8073, ISO 8327, ISO 8571, CCITT X.400 (1984, 1988), CCITT X.500, ISO 9595, ISO 9596
Reference [PTSV 93], p. 329	
ITEX Access Library	set of C++ classes that provide access to all of the functions of the editor (menu commands, keystrokes, etc.) from C++, and access to the parse tree of the test suite
ITEX C-Translator	automatic translation of an ATS in TTCN to an ETS, an executable program, in C
ITEX-DE Editor	TTCN test suite editor, compliance checking of test suites with the standard, comparison and merging of separate test suites, import and export tools for reading and writing TTCN.MP
STED	STED is a connection between the SDL design tool and ITEX, synchronization of test suites with SDL specifications of IUTs, automatic transfer of the static information (PCOs, data types and signals) from the SDL specification to the test suite, semiautomatic construction of test cases (signals can be sent into the specification and all signals that may be received will be entered in the test case), graphical support of the paths covered by a particular signal sent into the specification.
TeleLOGIC AB P.O. Box 4128 S-203 12 Malmoe Sweden	Tel : +46-40 17 47 00 Fax : +46-40 17 47 47 Email : info@telelogic.se
ITS	test operator support, monitoring, test report and diagnosis production
Reference [PTS 90], p. 369	
K1195	ISDN test tool
Reference [PTS 90], p. 370	
KATE	OSI protocol test system
Reference [PTS 90], pp. 215-229	

TABLE B.1 continued(3). Test tools

TEST TOOL	CATEGORY
OCTOPUS	Conformance Test Suite Development Environment, TTCN-GR/ ETS translator, TTCN-MP/ETS translator on top of SS7 TCAP, PICS-, PIXIT processing
Reference [PTSV 93], p. 331	
OSI Session Inter-Operation Verification Service	interoperability testing between the IUT and IBM OSI layers 1 to 5 Implementations
X.400 Inter-Operation Verification Service	interoperability testing between the IUT and IBM X.400 Implementations
Reference [PTS 90], p. 371	
OSITEST	test case generation / test execution / test result analysing system for
/400-88	message handling (X.400 series, 1984 and 1988 version)
/500	directory service (X.500 series)
/FAX4	facsimile group 4 class 1 (FAX4)
/FTAM	file transfer access and management (ISO 8571)
/NM	network management (X.700 series)
/ODA	testing of data stream conformance for ASN.1 (ISO 8824, ISO 8825), ODA (ISO 8613), and document application profiles (DAP) (Q112, FOD11, FOD26)
OSITOOL/ASSP1	abstract syntax and semantic processor 1, user configurable ASN.1 tool kit containing ASN.1 editor, encoder, decoder, semantic analyzer and configurator
OSITOOL/PICS-ED	PICS editor for the administration and production of PICS and PICS proformas. The input/output of the PICS editor is based on the results of ETSI-PT24V "Development of an open testing environment"
OSITOOL/TCM	test campaign management for the support of the administration process. The functionality includes test case selection, test suite parameterisation, PCTR and SCTR generation and SUT management.
OSITOOL/TTCN.AGR	TTCN.MP/GR converter and cross-reference production tool. This tool is able to produce TTCN.GR in plain ASCII format and cross-reference lists for each TTCN object contained within the TTCN.MP.
For non US: Danet GmbH Business Unit Telecommunications Technologies Roland Kuhn Tel: +49 6151 868-246 P.O.Box 10 04 03D-64204 Darmstadt Fax: +49 6151 868-131 Germany E-Mail: kuhn@danet.danet-gmbh.d400.de X.400: C=DE/A=D400/O= DANET-GMBH/P= DANET/ S= KUHN	

TABLE B.1 continued(4). Test tools

TEST TOOL	CATEGORY
For US: GSI Danet Inc. Hans-Ludwig Heil 1380 Old Freeport Road Pittsburgh, PA 15238 U.S.A.	 Tel: (412) 967-0834 Fax: (412) 963-6779 X.400: C=US/A=USA/ O=GSIDANET/P=GSI-DANET/S=HEIL
SDL Tool Set	SDL based tool set
Reference [PTS 90], p. 370	
SITE	testing environment , test execution manager for controlling parallel test components, editor, test case generator, TTCN.GR/MP converter, TTCN.MP compiler, monitor, automatic trace and report production. The test system is used in the ISDN, DECT and GSM protocol areas.
SIGOS Systemintegration GmbH Klingenhofstraße 50c D-90411 Nürnberg Germany	 Tel: +49 911 95 16 80 Fax: +49 911 51 44 11 E-Mail: mloehlein@sigos.de
SSC	TTCN static semantics checker
Reference [PTSV 93], p. 325	
TTCN Based Test System	
Reference [PTSV 93], p. 332	
Tekelec's Ch32-plus Conformance Testing Platform	protocol analysis and test execution for X.25, X.75, ISDN
Reference [PTSV 93], pp. 328-329	
UBC	protocol testing environment, Ferry based test system, UBC trace analyzer, UBC test case management system, UPC protocol tester
Reference [PTSV 93], pp. 323-324	
XRTLE	test suite development and test execution system
Reference [PTSIII 93], p. 462	

TABLE B.1 continued(5). Test tools

APTS/2	The Automated Protocol Test System/2 includes facilities for creating and editing abstract test suites using TTCN, defining test suite data for data structure specifications using the ASN.1 syntax, creating and editing PICS and PIXIT proformas, completing PICS and PIXIT proformas for a specific application, executing a test suite using specific PICS and PIXITs, and generating formal test reports. The PICS and PIXIT features of the tool are protocol independent, while the TTCN Editor and Execution functions require extensions (ASN.1 definitions of PDUs/ASPSs and drivers for the underlying service interface) for each of the protocols which they support. APTS/2 is an implementation of the remote test method described in the ISO standard. APTS/2 supports test suites written for the distributed or coordinated test methods; however, implementation of an upper tester is the responsibility of the user. APTS/2 is not intended to be used as a local tester. The APTS/2 user can produce: Abstract Test Suites, PICS and PIXIT proformas, completed PICS and PIXITs, conformance test logs and PCTRs. APTS/2 will run on an all models of the IBM (*) PS/2 (*) which contain an 80386 processor, or higher. To run the Execution component, the PS/2 must have a Microchannel Architecture bus and the appropriate communication adapter. APTS/2 runs as an application program in the OS/2 (*), Version 2.1, environment. It uses the IBM Presentation Manager (*) graphical support to manage its window driven user interface. (*) Trademark or Service Mark of International Business Machines Corporation in the United States or other countries.
Wilhelm Stoll IBM Deutschland Informationssysteme GmbH European Networking Center Vangerowstr. 18 69115 Heidelberg Germany	Tel.: (+49) 6221 59 4421 Fax: (+49) 6221 59 3300 E-mail: wstoll@vnet.ibm.com

ANNEX C. LIST OF ABBREVIATIONS

ACSE	association control service element
ASN.1	abstract syntax notation one
ASP	abstract service primitive
ATM	abstract test method
ATS	abstract test suite
BIT	basic interconnection test
BNF	Backus Naur/normal form
BRM	basic reference model
c, C	conditional capability or parameter
CCITT	comité consultatif international pour télégraphique et téléphonique (international consultative committee for telephone and telegraph)
CD	committee draft
CM	coordination message
CP	coordination point
CTMF	conformance testing methodology and framework
DIS	draft international standard
DP	draft proposal
ESTELLE	extended state transition language
ETS	executable test suite
F	fail
FDT	formal description technique
FIFO	first in first out
FMCT	formal methods in conformance testing
I	inconclusive
ICS	implementation conformance statement
IEC	international electrotechnical commission
IOT	interoperability testing (inofficial)
IPRL	ISP requirements list
IS	international standard
ISDN	integrated services digital network
ISO	international organization for standardization
ISP	international standardized profile

ISPICS	international standardized profile implementation conformance statement
ISPIXIT	international standardized profile implementation extra information for testing
ITU	international telecommunication union
ITU-T	ITU, telecommunication standardization sector
IUT	implementation under test
IXIT	implementation extra information for testing
JTC	joint technical committee
LT	lower tester
LTCF	lower tester control function
MHS	message handling systems
MOT	means of testing
MPyT	multi-party testing context
MTC	main test component
n, N	not implemented
n/a, N/A	not applicable
NE	negotiated exit
ODP	open distributed processing
OSE	open systens environments
OSI	open systems interconnection
P	pass
PATS	parameterized abstract test suite
PCO	point of control and observation
PCTR	protocol conformance test report
PDU	protocol data unit
PETS	parameterized executable test suite
PICS	protocol implementation conformance statement
PIXIT	protocol implementation extra information for testing
PSD	profile specification document
PSTS	profile specific test specification
PTC	parallel test component
PTS	profile test specification, parameterized test suite
RL	requirements list
SAP	service access point

SATS	selected abstract test suite
SC	subcommittee
SCS	system conformance statement
SCTR	system conformance test report
SETS	selected executable test suite
SPyT	single-party testing context
SUT	system under test
TC	test case (inofficial), technical committee
TCP	test coordination procedure
TMP	test management protocol
TM-PDU	test management PDU
TP	transaction processing, test purpose(s), transport protocol
TPDU	transport PDU
TR	technical report
TS	test suite (inofficial)
TSS	test suite structure
TTCN	tree and tabular combined notation
TTCN.GR	TTCN, graphical form
TTCN.MP	TTCN, machine processable form
UCS	universal coded character set
UPU	universal postal union
UT	upper tester
WG	working group
x, X	excluded capability or parameter
y, Y	implemented capability or parameter

REFERENCES

[BAER 94] U. Bär: OSI-Konformitätstests: *Validierung und qualitative Bewertung*, VDI Verlag, 1994

[BAGI 92] B. Baumgarten, A. Giessler: *OSI Testmethodik und TTCN*, GMD-Bericht Nr. 202, R. Oldenbourg, 1992

[BAUM 88] B. Baumgarten: *Structural Prerequisites for Unambiguous Conformance Testing,* 2nd ISIIS, INTAP,Tokyo, 1988, 283-290

[BAUM 94] B. Baumgarten: *Open Issues in Conformance Test Specification,* 7th Int. Workshop on Protocol Test Systems, Kogakuin University, Tokyo, 1994

[BEIZ 90] B. Beizer: *Software Testing Techniques,* Van Nostrand Reinhold 1990

[BGP 93] B. Baumgarten, A. Giessler, C. Paule: *The CAB Protocol – a TTCN Test Suite Generation Case Study*, Arbeitspapiere der GMD, Nr. 731, 1993

[EWOS 93] EWOS Technical Guide 025 Revision 1: *The TTCN Style Guide*, Brussels, 1993

[FMCT 94] ISO/IEC JTC 1/SC 21/WG 1: *Formal methods in Conformance Testing*, Working Document N 1317, 1994

[GOTZ 92] R. Gotzhein: *On Conformance in the Context of Open Systems,* 12th ICDCS, Yokohama, 1992, 236-243

[HALS 92] F. Halsall: *Data communications, computer networks and open systems*, 3rd Ed., Addison-Wesley, 1992

[HOGR 91] D. Hogrefe: *OSI formal specification case study: the Inres protocol and service*, TR IAM-91-012, 1991

[ISO 7498] ISO 7498, *Information Processing Systems – Open Systems Interconnection –Basic Reference Model,* 1984

(new version: ISO/IEC DIS 7498-1, *Information Technology – Open Systems Interconnection – Reference Model*, 1992)

[ISO 8824] ISO/IEC DIS 8824-1, *Information Technology – Open Systems Interconnection –Abstract Syntax Notation One (ASN.1) – Specification of Basic Notation*, 1992

[ISO 8825] ISO/IEC DIS 8825-1, *Information Technology – Open Systems Interconnection –Specification of ASN.1Encoding Rules – Part 1: Basic Encoding Rules (BER)*, 1992

[ISO 9646-1] ISO/IEC 9646-1, *Information Technology – Open Systems Interconnection –Conformance Testing Methodology and Framework, Part 1: General Concepts*, 1991 (2nd ed. expected 1994)

[ISO 9646-2] ISO/IEC 9646-2, *Information Technology – Open Systems Interconnection –Conformance Testing Methodology and Framework, Part 2: Abstract Test Suite Specification*, 1991 (2nd ed. expected 1994)

[ISO 9646-3] ISO/IEC 9646-3, *Information Technology – Open Systems Interconnection –Conformance Testing Methodology and Framework, Part 3: The Tree and Tabular Combined Notation*, 1992 (2nd ed. expected 1994)

[ISO 9646-3-1] ISO/IEC 9646-3 AM1: *Amendment 1: TTCN Extensions*, expected 1994

[ISO 9646-3-2] ISO/IEC 9646-3 DAM2: *Amendment 2: Further TTCN Extensions*, 1994

[ISO 9646-4] ISO/IEC 9646-4: *Information Technology – Open Systems Interconnection –Conformance Testing Methodology and Framework, Part 4: Test Realization*, 1991 (2nd ed. expected 1994)

[ISO 9646-5] ISO/IEC 9646-5: *Information Technology – Open Systems Interconnection –Conformance Testing Methodology and Framework, Part 5: Requirements on Test Laboratories and Clients for the Conformance Assessment Process*, 1991 (2nd ed. expected 1994)

[ISO 9646-6] ISO/IEC 9646-6: *Information Technology – Open Systems Interconnection –Conformance Testing Methodology and Framework, Part 6: Protocol Profile Test Specification*, (expected 1994)

[ISO 9646-7] ISO/IEC DIS 9646-7: *Information Technology – Open Systems Interconnection –Conformance Testing Methodology and Framework, Part 7: Implementation Conformance Statements*, (expected 1994)

[ISO 10000-1] ISO/IEC TR-10000-1: *Information Processing Systems – International Standardized Profiles – Part 1: Taxonomy Framework*, 1989

[ISO 10000-2] ISO/IEC TR-10000-2: *Information Processing Systems – International Standardized Profiles –Part 2: Taxonomy of Profiles*, 1989

[ISO 10646] ISO/IEC DIS 10646: *Information technology– Universal Coded Character Set (UCS)*, 1990

[JAAG 90] B.N. Jain, A.K. Agrawala: *Open systems interconnection – its architecture and protocols*, Elsevier, 1990

[MYER 91] G. J. Myers: *The Art of Software Testings,* Wiley, 1979

[PERR 83] W. E. Perry: *A Structured Approach to Systems Testing*, QED Information Sciences, Inc., Wellesley, USA, 1983

[PHAL 93] M. Phalippou: *The Limited Power of Testing*, Protocol Test Systems, V , North-Holland, 1993, 43-54

[PROST 93] Programme of Research on Open Systems Testing: *Report of the Study on Testing for Open Distributed Processing*, Architecture Projects Management Ltd., Cambridge, UK, 1993

[PTS 90] J.de Meer,L.Mackert,W.Effelsberg (eds.): *Protocol Test Systems,* North Holland, 1990

[PTSIII 91] I. Davidson, D.M. Litwack (eds.): *Protocol Test Systems,* III, North Holland, 1991

[PTSV 93] G. v. Bochmann, R. Dssouli, A. Das (eds.): *Protocol Test Systems,* V, North Holland, 1993

[ROSE 90] M.T. Rose: *The Open Book, A Practical Perspective on OSI*, Prentice Hall, 1990

[RUSS 89] D. Russell: *The Principles of Computer Networking,* Cambridge University Press, 1989

[SHOO 83] M.L. Shooman: *Software Engineering*, McGraw-Hill, 1983

[STAL 90] W. Stallings: *Handbook of Computer Communications Standards*, Vol.1., Howard W. Sams, 1990

[STEE 90] D. Steedman: *Abstract Syntax Notation One (ASN.1) – The Tutorial and Reference*, Technology Appraisals, 1990

[TRET 92] J.G. Tretmans: *A Formal Approach to Conformance Testing*, CIP - Gegevens Koninklijke Bibliotheek, Den Haag, 1992

[TURN 93] K.J. Turner (Ed.): *Using Formal Description Techniques*, Wiley, 1993

INDEX